Praise for *The War on Ivermectin*

"Dr. Pierre Kory is one of the heroic figures of our time; a courageous physician who sacrificed career, reputation, friendships, status, and livelihood for the health of his patients and humanity."
—Robert F. Kennedy Jr., founder of Children's Health Defense and author of *The Real Anthony Fauci*

"We will never know how many people would be alive today had the medical establishment taken their blinders off, opened their minds, and listened to doctors who had the courage and compassion to actually treat COVID patients. Instead, heroes like Dr. Pierre Kory were ridiculed, vilified, censored, and professionally cancelled. Dr. Kory's must-read book is a fast paced, engaging firsthand account of what went terribly wrong in our medical response to the pandemic. We must learn from his experience and use his insights to repair our health system that has been corrupted by Big Pharma and the hubris of those in power."
—Senator Ron Johnson

"Dr. Pierre Kory's book, *The War on Ivermectin*, is an exceptional contribution to the literature that has emerged from the COVID-19 pandemic. It provides insight into little-known facts about the pandemic and is written in a captivating fashion that mimics Dr. Kory's entertaining manner of speaking. The book highlights how widespread adoption of Dr. Kory's work could have saved countless lives. By reading this book, you will deepen your understanding of the fraudulent practices associated with discrediting ivermectin and discover why it is essential to have access to this safe and effective therapy for potential future parasite or severe viral infections."
—Dr. Joseph Mercola, founder of the world's most visited natural medicine site – mercola.com

"In 2020, when Dr. Pierre Kory told America that steroids were efficacious for COVID-19 pneumonia, every regulatory authority in the world said not to use them. Within months, the world followed Kory. Later that year he appeared on the Senate floor and advocated for every acutely ill patient to get a chance to survive COVID-19 with a "miracle" drug, ivermectin. He was at the epicenter of what became the greatest tragedy of therapeutic nihilism in the history of our country. America lost over a million souls to the virus and virtually all of them were denied this safe and effective compound. *The War on Ivermectin* can only be told by Kory, a man with a drug the biopharmaceutical complex tried to destroy because they knew it would end the pandemic. It is a powerful vision of courage and strength overcoming adversity to save as many as possible. Against all odds, Dr. Kory and his band of brothers, the Front Line COVID-19 Critical Care Alliance, fought with the alacrity of a rebel squadron against a ruthless, limitless empire bent on malice for the world. This book is a must-read for this time and all time. Kory and his indomitable spirit, the FLCCC, and every doctor, nurse, and patient blessed by the wonder drug ivermectin touched the currency of a dark war the world cannot possibly understand."

—**Peter A. McCullough, MD, MPH, author of *The Courage to Face COVID-19: Preventing Hospitalization and Death While Battling the Bio-Pharmaceutical Complex***

"Dr. Pierre Kory's testimony before the US State Senate in December 2020 changed my life. As a medical doctor and external consultant to the World Health Organization, it amazed me that an experienced US doctor should have to beg politicians for permission to use little old ivermectin—a cheap and essential Nobel prize-winning medicine, on the WHO's list of essential medicines, with an immaculate safety record. Ivermectin is the key to revealing all the lies we've been told about COVID. *The War on Ivermectin* is a tale of corruption, censorship, and criminal intent, and no one was more on the front line during COVID and the ivermectin wars than Dr. Pierre Kory and Dr. Paul Marik. This personal account by Pierre about how the evidence on ivermectin was attacked and buried for three years by an anti-human Big Pharma cabal is so rich and detailed, it will enthrall the most knowledgeable on the matter, and blow the minds of those new to it. The genie is now out of the bottle. The war on ivermectin is a war on humanity. In my opinion, the most important thing that Pierre's written experience reveals is that when science and reason fail, and when the enemy is all-powerful, there is no better way to navigate life than with our hearts. That said, there is no bigger heart on this battlefield than that of dear Dr. Pierre Kory who may have lost a few jobs in the COVID years but has gained the love and respect of millions."

—**Tess Lawrie, MBBCh, PhD, director of The Evidence-Based Medicine Consultancy and founder of the World Council for Health**

"Dr. Pierre Kory has been a tireless truth-seeker and advocate for early COVID treatment from the beginning of the pandemic. In *The War on Ivermectin*, he powerfully documents the deep-seated corruption and relentless propaganda that led to the greatest humanitarian catastrophe in history. As important as it is to acknowledge the lives saved by his advocacy, it is also critical to recognize the carnage that continues to this day despite the herculean efforts of Dr. Kory and a small but steadfast army of warriors like him. In deftly exposing the systemic corruption in our medical systems and media, he has solidified his place among COVID pandemic heroes. I applaud his courage and tenacity, and I highly recommend this important book."
—Aseem Malhotra, MBChB, cardiologist and author of *The 21-Day Immunity Plan*

"It's now three years after the start of the pandemic and we know from over ninety-five independent studies that ivermectin is one of the most effective drugs for treating COVID-19. The one person who deserves the most credit for bringing this drug to the public's attention very early on is Dr. Pierre Kory. This book chronicles in great detail what happens to a courageous doctor who speaks the truth and calls for 'letting doctors be doctors.' I have the utmost respect for Dr. Kory, and I'm sure you will too after you read his story."
— Steve Kirsch, founder of COVID-19 Early Treatment Fund (CETF) and Vaccine Safety Research Foundation (VSRF)

"A highly readable and fascinating memoir about the unfortunate demise of evidence-based medicine into what Dr. Kory calls 'evidence-based mania.' *The War on Ivermectin* is an insider's account about how, starting in March of 2020, best medical practices were displaced by corruption, blindness, and corporate greed. Highly recommended."
—Jennifer Margulis, PhD, investigative journalist and award-winning author of *Your Baby, Your Way* and coauthor with Dr. Paul Thomas of the #1 Amazon bestselling book, *The Vaccine-Friendly Plan*

"This book should be made into a sinister action movie! Dr. Kory has elegantly captured a gripping drama where ivermectin is the main character in a dark and complicated plot line. As the story unfolds, a larger pattern of deceit and callous carelessness is revealed that will forever change your view of how the so-called "healthcare" system actually works. Although the movie ends as a cliffhanger (will the protagonist eventually prevail?), it points the way to reclaiming your own agency in your health and medical outcomes, and provides much needed hope which leaps from these pages as you are introduced to the many doctors who truly care, actually follow the evidence, and are using their time to make this world a better place."

—Chris Martenson, PhD, founder and CEO of Peak Prosperity and author of *The Crash Course*

"The COVID-19 pandemic changed the world. Among the most profound changes was a coup by public health bureaucrats against doctors and the traditional practice of medicine. Medical professionals and scientists were told how to think, and those who bristled at the new order, or at the broken logic it dispensed, were ridiculed, censured, and cast out. There was to be one and only one answer to the pandemic—a pair of unprecedented, barely tested therapeutics portrayed to the public as "safe and effective vaccines." The only hitch in that plan was that these novel therapeutics couldn't possibly get FDA approval fast enough. Instead, these brand-new shots would need to get Emergency Use Authorization, which, because of the danger of experimenting on the general public, required there be no available alternative. That requirement is why an all-out war has been waged on ivermectin, a drug that safely and effectively treats and prevents COVID but isn't worth money because it's no longer under patent. If the truth of ivermectin were ever recognized, it would bring the full criminality of global COVID policy into stark relief. *The War on Ivermectin* tells this shocking story in vivid detail from the vantage point of Dr. Pierre Kory, a heroic and highly decorated ICU specialist who spent every minute fighting on the front lines of this dirty covert war waged by Pharma and its governmental captives against doctors, patients and all the citizens of Earth."

—Bret Weinstein, PhD, evolutionary biologist, cohost of the DarkHorse Podcast, and coauthor of *A Hunter-Gatherer's Guide to the 21st Century*

THE WAR ON IVERMECTIN

THE MEDICINE THAT SAVED MILLIONS AND COULD HAVE ENDED THE PANDEMIC

DR. PIERRE KORY
WITH JENNA MCCARTHY

FOREWORD BY DEL BIGTREE

Skyhorse Publishing

Skyhorse Publishing books may be purchased in bulk at special discounts for sales promotion, corporate gifts, fund-raising, or educational purposes. Special editions can also be created to specifications. For details, contact the Special Sales Department, Skyhorse Publishing, 307 West 36th Street, 11th Floor, New York, NY 10018 or info@skyhorsepublishing.com

Skyhorse® and Skyhorse Publishing® are registered trademarks of Skyhorse Publishing, Inc.®, a Delaware corporation.

Visit our website at www.skyhorsepublishing.com.

10 9 8 7 6 5 4 3 2 1

Library of Congress Cataloging-in-Publication Data is available on file.

Hardcover ISBN: 978-1-5107-7386-8
eBook ISBN: 978-1-5107-7387-5

Cover design by Brian Peterson

Printed in the United States of America

Americas Front line Doctors
AFLDS

Front line Covid Critical
care   FLCCC

Dedication

I could not have written this book without the unwavering patience,
sacrifice, and love of my beautiful wife Amy and incredible daughters,
Ella, Eve, and Violet.
You supported and championed me as I allowed my fight against
Covid to consume my every waking moment.
Although they both took "Poppy" away, that battle and
this book were always for you.

Contents

Foreword

I am not a doctor, but I made them famous on TV. As one of the Emmy-winning producers of the CBS talk show *The Doctors*, I was tasked with searching the world for groundbreaking advancements in science and medicine and turning those stories into entertainment for daytime television viewers across America. In the six years I worked on the show, I did everything from filming an anatomical woman being turned into an anatomical man to hosting a raucous debate between the head toxicologist behind Monsanto's herbicide Round-Up, and a renowned anti-Monsanto activist.

I have filmed in the OR while surgeons performed life-saving miracles before my eyes, and I have documented people submitting their bodies to doctors I wouldn't trust with my lawn mower. When I am asked what the biggest takeaway was from my years working on *The Doctors*, my answer is always the same. We, as lay people, think that medicine is constantly advancing at a fervent pace and that the best products and techniques are rising quickly to the top. We have faith that once a new approach has proven to be effective, it is shared amongst all the trusted doctors in the specialties of interest. Nothing could be further from the truth.

Medicine is one of the slowest evolving systems in the world. It is replete with massive egos that are incapable of embracing an idea that is not their own. And in my experience, the more earth-shattering and brilliant a new intervention, technology, or technique is, the more likely they are to cause the inventor's license to be revoked.

There are those who may think that the attack on genius is a modern phenomenon. But even a cursory browse through the annals of science will reveal that breaking away from consensus, which is the only way that science ever advances, is rarely good for a person's career. We all know that the earth is not the center of the universe, but when Galileo discovered this astronomical truth, he wasn't celebrated. He was accused of being a heretic and put under house arrest. No doctor in the world today would dare to move from one surgery to the next without washing their hands in between, but when Dr. Ignaz Semmelweis discovered the vital importance of hand washing, he wasn't celebrated by the medical establishment. He was ridiculed and admonished, ultimately dying an unceremonious death in an insane asylum. One would hope that after centuries of deplorable assaults by the scientific mob, a more dependable pathway for powerful ingenuity would have been constructed. But my work as a medical journalist has shown me that "genius shaming" is worse now than ever before.

At the same time that we ponder all of the current foundational principles of science that barely survived the tar and featherings, floggings, and beheadings by the establishment, it may be of even greater importance to focus our critical eye on all of the wretched interventions that have lighted softly upon the loving arms of scientific consensus without an inkling of resistance. In the 1960s, thalidomide was touted as a miracle drug for insomnia and morning sickness until enough babies were born without legs and arms. Vioxx was the answer to arthritis until a multi-billion-dollar lawsuit over tens of thousands of deaths revealed that Merck had always known the drug would cause heart attacks. How many millions of people developed cancer from having Johnson & Johnson baby powder sprinkled on their bottoms as babies only to find out in a multi-billion-dollar lawsuit that for the fifty years the product was on the market, the company always knew there was no way to avoid contamination by asbestos when mining for talc? Too often, we read these headlines in our newspapers, and we are enraged. But who is to blame? The manufacturer who made the product? The doctor who recommended the product? Or the government regulatory agency that promised us the product was safe?

It was actually my investigation into the safety of vaccines that caused me to leave my job at *The Doctors*. In 2015, I got a tip from a trusted source that there was a whistleblower inside the CDC that had 10,000 documents that were the evidence of his claim that the CDC had committed scientific fraud on a study evaluating the MMR vaccine and autism. This study should have revealed to the world that there appeared to be a correlation between the vaccine and autism. Instead, the CDC kicked half of the kids off the study and manipulated data tables to erase the damning results. When I pitched the story to my executive producers, they refused to let me investigate it because it would not make the CDC or our pharmaceutical sponsors happy. So I left my job and went on to make the documentary *Vaxxed: From Cover-Up to Catastrophe.*

Most people have no idea the power that the pharmaceutical industrial complex has over their television and, more specifically, the news that they watch. It has been reported that up to 70 percent of television advertising, especially during an election year, is purchased by Pharma. But you don't need a spy or a whistleblower to prove this. Just count the number of ads for pharmaceutical products during every commercial break on your next evening of television viewing. Ultimately, those advertisers are who are paying the news anchor you trust to give you the news. Do you know of any CEOs who knowingly sign paychecks to employees who undermine confidence in their products? What do you think would happen if a news reporter warned people about the dangers of a product made by a company that just paid for them to be on the air?

In 2017, after spending a year on tour with *Vaxxed*, I launched the nonprofit Informed Consent Action Network (ICAN). I also created a weekly news program through ICAN called *The HighWire.* Our dream was to create a news organization without sponsorship so we'd be free to investigate anything we wanted, whenever we wanted, without fear of losing our jobs. One of our primary investigations was vaccines, which are protected from lawsuits in a one-of-a-kind sweetheart deal signed by our government in 1986, so we enlisted the services of an attorney named Aaron Siri, who has helped us reveal government corruption by winning lawsuits against regulatory agencies including HHS, NIH, CDC, and the FDA.

Perhaps the most insidious of all the problems with mainstream medicine and science is the incestuous relationship it has with the US government. It is now well established that the pharmaceutical industry is the number one most powerful lobby in Washington. When speaking to audiences, I have often said, "Pharma is outspending the oil and gas lobby two to one. If we fight unending wars in the middle east for oil and gas, what do you think our government is going to do for Pharma who is giving them twice as much money?" Is it possible that this investment can explain why there is a revolving door between our regulatory agencies and the pharmaceutical industry? When we watch the commissioner of the FDA, Scott Gottlieb, leave his post as the watchdog responsible for ensuring pharmaceutical products are safe to take a position on the board of Pfizer, are we allowed to ask, "Would he have gotten that position if he had blocked a Pfizer product from being approved because it was unsafe?"

It is clear to anyone who is paying attention that regulatory capture is likely the biggest issue we are facing in the world. If the CDC and FDA are being run by Pfizer and Monsanto executives who will return to multimillion-dollar salaries with these companies after their tenure with the government—the EPA is being run by Exxon and BP executives, and the FAA is being run by Boeing and Lockheed Martin executives—what confidence can we have that the products that they approve are selected because of their benefit to the people and not their benefit to the industries that got them their jobs?

By the time COVID made landfall in America, my team at *The HighWire* was ready for it. We knew that the government's regulatory agencies would push a vaccine because the agenda was obvious to us before it started. We knew that the media would supercharge the hysteria around the virus because that's what they taught us to do when we worked in television. We knew they would try to force everyone into compliance because there is no other reason a cold virus with a death rate similar to the flu would be touted as the greatest medical calamity of our lifetime. What we did not know is that one of the cheapest and safest drugs in the world would so vividly expose the corruption that threatens the future of science and medicine today.

The War on Ivermectin is not about a war on a miraculous little pill, it is about a war on humanity. Once you understand what happened here, you will understand what we need to do to save our world.

Del Bigtree
Host of The HighWire
CEO of Informed Consent Action Network

PART ONE

GEARING UP FOR BATTLE

CHAPTER ONE

Before the Beginning

Here's to the crazy ones. The misfits. The rebels. The trouble-makers. The round pegs in the square holes. The ones who see things differently. They're not fond of rules. And they have no respect for the status quo. You can quote them, disagree with them, glorify or vilify them. About the only thing you can't do is ignore them. Because they change things. They push the human race forward. And while some may see them as the crazy ones, we see genius. Because the people who are crazy enough to think they can change the world are the ones who do.

—Steve Jobs

I do quite a bit of public speaking these days, and part of my schtick has become somewhat of an "ode to the old Pierre." When I say old, of course, I mean pre-Covid.

Old Pierre believed that the elite, esteemed medical journals represented the best of scientific thought and study. *The Lancet* or the *New England Journal of Medicine* said so? It was settled then. Old Pierre religiously read the *New York Times* from cover to cover, because it was the paper of record; the arbiter of truth. If you wanted to know what was really going on, you read the *Times*. Period. He voted for Biden (although in his defense, he wasn't exactly a fan and never put a BIDEN-HARRIS ring around any of his social media profile photos), trusted the government (I know!), and actually believed that public health agencies were committed to safeguarding and improving . . . wait for it . . . public health. He knew—*knew, I tell you!*—that vitamins were a scam and that hospitals were life-saving centers of care, compassion, and excellence. Old Pierre dutifully lined up for his own annual flu shot and followed the childhood immunization schedule to the letter with his three daughters.

He was a clueless sonofabitch.

Nobody, least of all me, could have predicted the insane series of events, discoveries, and decisions that would transform him (me) into the wildly different doctor—and man—that I am today.

But here we are.

So this is my story. What started as a daily brain dump, a place to record the happenings and heartbreaks occurring at work and at home, slowly morphed into this crazy peek into a decidedly broken medical system. I set out to understand and expose what was happening with repurposed drugs, ivermectin specifically. By October of 2020, we had identified an inexpensive, safe, widely available medication that was showing tremendous potential not just as a treatment for Covid but also as a preventative. As the weeks and months wore on, the data supporting its safety and efficacy were astounding. And yet the backlash against it was swift and furious. Positive studies were overturned and retracted. Negative studies appeared out of thin air. Around the world it was quietly being used to tremendous, almost impossible success, and yet doctors were punished for prescribing it, pharmacies refused to fill valid prescriptions for it, and the media would only touch it to call it "the horse dewormer." To a physician fighting on the front lines of this battle, this systematic smear campaign was unfathomable.

I soon discovered that the corruption and deceit were hardly limited to the pharmaceutical space. The entire medical industrial complex—including our governmental and international regulatory agencies, Big Pharma, public and private health care systems and hospital networks, medical schools and their journals, and at least one centi-billionaire "philanthropath"—had been collectively captured. According to Wikipedia (which I don't often use as a reference source, incidentally, but their explanation was most succinct), "When regulatory capture occurs, a special interest is prioritized over the general interests of the public, leading to a net loss for society."[1]

You can say that again.

At the risk of sounding arrogant or self-congratulatory, when it came to Covid, I got a lot of things right from the beginning. So often and so overwhelmingly, in fact, that I was dubbed "Lucky Pierre," first by the editor of the *New England Journal of Medicine* in a magazine interview, and then by my colleagues and friends. I want to acknowledge here, up front, that I ascribe much of that consistent, almost implausible "rightness" to this: practically from day one, I was part of a group of highly credible, extensively experienced professors, scientists, and clinicians who were deeply studied on nearly every aspect of medicine even remotely related to Covid. We shared a spirit and a purpose well before we had a name (the Front Line Covid Critical Care Alliance, or FLCCC), a website, or a nonprofit designation.

The whole is always greater than the sum of its parts, and that is exponentially true with the FLCCC. After all, we're the misfits, the troublemakers, the round pegs in the square holes. We're the ones standing up to the system; the child watching the bare-assed Emperor parade down the street who just can't hold his tongue.

"But he hasn't got anything on," we've been shouting. At first, people pointed and laughed at us and called us names, but we didn't care. That fat bastard was naked, and nothing could make us see or think otherwise! And do you know what? People are starting to catch on. More and more, some might say in droves, they're seeing what we see and have seen for a few years now.

That is the power and spirit of the collaboration and camaraderie behind the FLCCC. From the beginning, we were bound by mutual passion and respect, and committed to uncovering and speaking the truth—no matter how difficult or isolating that proved to be.

So yeah, we've gotten a lot right. It turns out, that's actually not so hard to do when you're surrounded by greatness and your hearts are in the right place.

CHAPTER TWO

Foreshadowing

"Feck, feck, feck," Paul yelled. (That's how Paul sounds when he curses in his South African accent). "It's negative!"

"What?" I asked.

"It's fecking negative!"

"What do you mean? How the hell can it be negative?"

"I don't know," Paul bellowed. "I just got the paper and I'm already at the airport. Those bastards purposely didn't send it to me on time!"

"I need to see it," I insisted.

"I'm not supposed to share it, it's embargoed until Thursday."

"Screw that, Paul!" Now I was yelling, too. "Send it to me. I have to see it. It's negative for everything?"

"Everything."

"Even the time on vasopressors?"

"Yes."

"Mortality and length of stay?"

"Yes."

"Paul, they did something stupid. We *know* it works, there's no way the study could be negative. It's not possible!"

Paul's reaction was more than justified. He had just learned that the world's first large, prospective, multi-center, double-blind, randomized controlled trial on the impacts of high-dose intravenous vitamin C (IVC) in septic shock was negative—meaning that the trial concluded it had no impact on any important outcome in the patients treated.

Paul and I both knew that this was utter bullshit.

Paul Marik isn't just an accomplished physician and researcher, or a former tenured professor of medicine, or the author of hundreds of peer-reviewed journal articles and four critical care textbooks. Paul is also an IVC expert, renowned for developing a lifesaving protocol used to treat sepsis, a condition that causes more than 250,000 deaths per year in the US alone, and according to research published in 2020, is the leading cause of death worldwide—above cancer and cardiovascular disease.[1]

That conversation happened on January 16, 2020, before the words "Covid-19" and "pandemic" were staples in the global vocabulary. Paul was on his way to an international conference in Belfast, Ireland, called Critical Care Reviews, which would feature an unveiling of the results of the previous year's most important trials in ICU medicine. There were a lot of eyes on this conference because the medical world was anxiously awaiting the results of the first randomized controlled trial of IVC in sepsis, moderated by the great Paul Marik.

It was also going to be the first time Paul and I met in person after spending countless hours on the phone as friends and colleagues over the prior two years. We had no idea that this conference would be consequential for entirely different reasons than we had anticipated.

It has taken the painful clarity of hindsight to realize how naive and ignorant we were then, at least in regard to the academic medical system—one we had been practicing, researching, and teaching in for decades. We were wholly unaware that the events about to unfold over the next two days would be the start of what has turned into a relentless three-year battle with a medical system we've since discovered has been completely corrupted and captured by the pharmaceutical industry.

I would argue that Paul should be more embarrassed about his ignorance at that time than I, as he had long been considered a pioneer in medicine. He was trained in Critical Care, Neurocritical Care, Pharmacology, Internal Medicine, Anesthesia, Nutrition, and Tropical Medicine and Hygiene, and was a tenured Professor of Medicine and Chief of the Division of Pulmonary and Critical Care Medicine at Eastern Virginia Medical School (EVMS) in Norfolk, Virginia. He has published over 500 peer-reviewed journal articles, written eighty book chapters, authored four critical care books, and has been cited over 43,000 times in peer-reviewed publications. In medicine, scientists are given a ranking of their impact to their field by way of something called the h-index; Paul has an h-index of 110. For reference, a typical h-index for a professor ranges from 12 to 24, and most Nobel Prize winners score 30 or above. What I'm saying is, Paul is a fecking force. Further, he has

delivered over 350 lectures at international conferences and visiting profes-
sorships and won numerous awards, including the National Teacher of the
Year award by the American College of Physicians in 2017.

At the time of the Belfast conference, I was a mere twelve years out of
training. Paul had decades of practice under his belt. I was—and still am—a
rookie compared to him. That conversation would mark the beginning of
our journey together, not only because of its timing, but more so in terms of
how it so powerfully pitted us against the broken academic medical system.

All the corruption, the disinformation, the fraudulent trials, and the
editorial attacks on IVC would start with this day. Of course, we did not yet
know about the root cause of it all: a systematic corruption in the medical
sciences by an industry that has been targeting repurposed drugs, vitamins,
and alternative therapies for decades.

When we read the study, we quickly picked up the investigators' mistake: they had given Paul's vitamin C protocol too late into the disease.
Our reaction to this discovery was more disappointment than anger, as we
simply and naively assumed that the trialists designed the study out of ignorance of the importance of timing of interventions in critical care. It almost
makes me laugh that that was my actual interpretation of the outcome.
These were world renowned critical care experts, mind you, and they had
allegedly designed a trial where patients in septic shock would not receive
the study treatment for up to as much as thirty hours at our best estimate.
These elite trialists had ignored the core concept of critical illness resuscitation,
which is the importance of "the golden hour," meaning with every minute
or hour that goes by before instituting effective therapy, the probability of
improvement rapidly diminishes.

I ask you: How do critical care physicians, researchers, and academics
running a potentially ground-breaking trial "forget" such a fundamental
concept? How does something so significant in the setup of a trial get over-
looked? What experienced ICU physician would *ever* make this kind of
mistake? I can't think of one. In retrospect, the whole debacle parallels the
past few years' global amnesia regarding the protective effects of natural
immunity while embarking on a mad pursuit to vaccinate the world against
a highly mutagenic virus that they've likely recovered from.

It was our later experience as "advocates" (ugh) of the medicine called
ivermectin in the early treatment and prevention of Covid that would finally
make us realize that the categorical destruction of a proven therapeutic was
not borne of stupidity. The people in charge knew EXACTLY what they
were doing . . . and always had. We should have figured that out already

with what was happening to IVC. But we didn't. We were slow on the uptake, or rather, like almost all physicians working in what I now call "The System," simply too trusting of its institutions and leaders and the objectivity of the "science" published in the top medical journals.

Our interest in the use of intravenous vitamin C in sepsis is what brought Paul and me together as colleagues and then friends, a friendship that began after I wrote an editorial that was published in a major medical journal called *Chest* and had been strongly influenced by Paul's published work.

I was shocked when he wrote me an email congratulating me on its publication. Little old me got a personal note from the great Paul Marik? He also attached an important paper that he thought I should have discussed and referenced. (In my defense, that paper had been published after I had submitted my editorial.)

Like many others in my field, I was an avid admirer of Paul. He was an intellectual giant in critical care medicine, the embodiment of a thought leader. What's funny is that so many so-called "thought leaders" are not leaders at all, but rather status quo-supporting, orthodoxy-upholding doctors with positions of authority or profound pharmaceutical or agency influence. They lead thoughts, alright . . . the ones they're guided, bribed, or forced to lead. Conversely, Paul's lectures at major academic conferences were always overflowing as his research and insights often led to conclusions that completely opposed prevailing orthodoxy and standards of care in the ICU. (See chapter four for a thorough probe into Paul's career exploits.)

More important than the fact that Paul often argued against the prevailing guidelines issued by the professional academic societies is that his data, analyses, and conclusions were nearly always impossible to rebut logically or scientifically. Time and again, Paul would show that the standard of care was not based on correct scientific data or an accurate understanding of the underlying pathophysiology of the disease or treatment. He has a gift for compiling and analyzing evidence and presenting it in such a way that is both compelling and humbling. Yet time and again, the academic societies were neither compelled nor humbled.

After receiving his email in 2016, I didn't write back to Paul for almost eighteen months. I kept his note bolded in my inbox waiting for a time when I felt I could reply, but the truth is I just couldn't. I was consumed in a health crisis involving one of my three daughters. Following a severe streptococcal infection, she had developed a serious, acute neuropsychiatric syndrome called PANDAS (Pediatric Autoimmune Neuropsychiatric Disorders Associated with Streptococcal Infections) at the time but now known as PANS (Pediatric

Acute-onset Neuropsychiatric Syndrome). She suffered intense neurologic symptoms that were traumatic for her to experience and unbearable to witness. Worst of all, they had suddenly appeared in a beautiful, social, happy, neurodevelopmentally normal, and highly intelligent child.

Her unexplained suffering was excruciating. Further, her symptoms were accompanied by a debilitating separation anxiety from my wife, Amy. This was especially challenging considering that Amy is also a pulmonary and critical care doctor who sub-specializes in a category of rare and difficult to treat disorders collectively known as interstitial lung disease (ILD). Although Amy was on a significant leave from work to care for our daughter, she tried to keep her one clinic day on Wednesdays.

Wednesdays were brutal.

During those years, I was on edge from the moment my daughter woke up to the time we could somehow get her to sleep at night. I was deteriorating physically and psychologically from stress and lack of self-care. My wife, on the other hand, was an absolute rock. When I acted out, Amy kept her cool. Still, the ordeal took an undeniable toll on us both.

Our daughter's illness consumed our days, our thoughts, and all of our intellectual and physical energies. In the span of just a few months, we had seventeen different encounters across emergency room physicians, pediatricians, neurologists, psychologists, and psychiatrists. One of the latter ultimately diagnosed my daughter with functional neurological disorder (FND), which is classified as a mental health condition. This is when I realized these "experts" knew nothing at all about this syndrome. The parallels to what is happening now to those with Covid vaccine injury syndromes—many of whom receive the same damning diagnosis as my daughter did—is beyond disturbing.

Our family PANS/PANDAS crisis was the first traumatic battle I had with the health system and one I will write about again at some point in the future. Compounding the PTSD is the belated knowledge that PANS/PANDAS exploded in frequency and severity at the same time the childhood vaccine schedule exploded in the late 1980s. I didn't know at the time that my family was being destroyed by a vaccine-associated disease.

Words cannot explain how profoundly disturbing that eventual realization was.

Fortunately, after months of delay in diagnosis and treatment, we finally found a brilliant pediatric neurologist. Despite her colleagues' misgivings and even condemnation for being willing to treat a disease that "doesn't exist," she was able to return my daughter to her completely normal neurological and social functioning. Not that recovery was fleeting or easy; it took months in an ICU with a combination of aggressive treatments including high-dose corticosteroids, plasmapheresis, intravenous immunoglobulins, and a B-cell depleting cancer agent.

So, to say that Covid is not my first scrimmage in battling modern medicine is a gross understatement. As a result of that experience, I got heavily involved with and became a board member of the incredible nonprofit now called the Neuroimmune Foundation[2] (formerly The Foundation for Children with Neuroimmune Disorders). Our mission was simple: to increase awareness and diagnosis of the disease, educate providers on treatment options, and fund research into better understanding the biological causes, diagnostic measures, and therapies. The founder and executive director of the foundation, Anna Conkey, became a close friend, colleague, and confidant, and I owe her the world for her help and support during that difficult time and throughout my Covid challenges. I love Anna and consider her a truly remarkable human being. (If you are able, please donate to neuroimmune.org.)

For the next several years, I moderated numerous webinars, lecture series, and symposia for the Neuroimmune Foundation, interacting with clinicians, scientists, and researchers in the disease. But in June 2021, after moderating that year's annual conference, Anna received complaints that I was "too controversial" to be associated with the foundation due to my public advocacy in Covid and with the FLCCC. I no longer host those educational events.

If my PANS battle was my first meaningful clash with academic medicine, my second was my experience teaching, treating, and researching the use of IVC in sepsis with Paul Marik. The IVC in septic shock story started when Paul began incorporating it into his treatment of such patients in 2016 based on a review and critical analysis of a few small studies showing absolutely dramatic reductions in mortality. He then published his experience with the first 47 patients he treated by comparing their outcomes with 47 patients selected from the prior year that were "propensity matched" to the IVC-treated patients in terms of age as well as cause and severity of sepsis. He reported that in IVC-treated patients, only 8 percent died, while in matched patients not receiving IVC, 40 percent had died. Let that sink in for a second.

There are very few medical interventions that lead to such a profound reduction in mortality. One way in which we measure the potency of an intervention in medicine is via use of a measure known as the Number Needed to Treat (NNT) to save a life (or prevent a stroke or illness). Paul's study found an NNT of 3.1, which meant that for every three patients he treated with IVC, one life *that would otherwise likely end* would be saved. A more disturbing way of putting it is that, for every three patients denied IVC in early sepsis, one would die unnecessarily. Let *that* sink in for a second.

Note that the most powerful intervention in medicine is the use of a defibrillator in someone whose heart has stopped. That intervention has an NNT of 2.5. Paul had discovered a therapy nearly as powerful in terms of its life-saving properties.

Paul published his study in the prominent journal *Chest* in 2017. I should note that his "protocol" was not just centered around the use of IVC but also included IV corticosteroids and IV thiamine. Here's the crazy part: When the paper and its results were published, I actually dismissed it as "too good to be true." There was no way that IVC could have that effect. It was unheard of. I had never come across any therapy that reduced mortality in the critically ill so profoundly. It just wasn't possible. Even if it *was* from the great Paul Marik.

Please keep in mind this is pre-Covid Pierre we're talking about here; the *New York Times*–blind, brainwashed disciple of The System. To that end, another reason I ignored Paul's study is that I had a keen and longstanding disdain for vitamins as any sort of therapeutic in acute illness. With a tiny, grudging exception for some chronic conditions, I considered the vitamin industry to be a billion-dollar scam that preyed on people who didn't need or benefit from them. As a system physician, I had been drowning in "negative" vitamin trials published in the most prestigious medical journals in the world for years; vitamin D trials for all sorts of illnesses and cancers, as well as vitamin E, A, and C in everything from immune disorders to cardiovascular disease. *Enough with the stupid vitamins,* I thought, *they're a total scam! The science is clear, it's right here in this esteemed journal.* (See where this book is going?) What I didn't know was that the difference between oral vitamin C and intravenous vitamin C is like the difference between a pistol and a machine gun.

A third reason I paid little heed to Paul's paper is that it generated quite a bit of media buzz. His hospital's press office allowed television stations to interview his nurses, who all but called the protocol *miraculous*. It was such a weird, unprecedented way to disseminate knowledge of a scientific breakthrough that I suppose I found it unprofessional, or at least unbecoming. I had never heard of a TV station interviewing nurses about some miracle therapy based on what I foolishly thought was a low-quality study. Of course, I now know such studies are highly valid, and their conclusions, especially when so large, are irrefutable.

So, for a bunch of ill-conceived reasons, I ignored Paul and his little study. I didn't know anyone at the University of Wisconsin who was using it, and I wasn't in the market for any snake oil.

Fast forward about a year. It was early 2018 (so still pre-Covid) and I was the director of the Medical ICU at the University of Wisconsin and the Chief of the Medical Critical Care Service. I was having a brutal week. My primary Medical ICU was so slammed with patients that they overflowed to other ICUs (Cardiac, Neurosurgical, Surgical).

Not-so-fun fact: My ICU mortality at UW was about 8–15 percent of patients admitted to me. That was the average. On a really bad week, I might see as many as 20–25 percent of my patients die. Feeling helpless and following my long-held principle in ICU medicine, "if what you're doing isn't working, *change what you are doing*," I decided, what the hell, why not try Marik's stupid IV vitamin cocktail? (Sorry, Paul.) I had nothing to lose.

The first patient I tried it on was a man decompensating from severe septic shock. He was already in advanced multi-organ failure, but his understandably distressed family was begging me to do everything I could. Unfortunately, despite initiating Paul's protocol, he died later that day. I was unsurprised by this as I already knew that nothing worked in actively dying patients. Still, it seemed totally harmless (which it was) given it consisted of a couple of IV vitamins and a corticosteroid (the latter of which was already part of my practice), and morally it felt better than doing nothing at all, so I figured I'd try it again.

The second patient was a female with necrotizing fasciitis—a deadly bacterial infection known as flesh-eating disease. I knew I needed a surgeon, so I consulted my friend and colleague at the time, Dr. Hee Soo Jung, who rapidly got his team together to take the patient to the OR. There they would do a debridement of her abdominal soft tissues which were red, hot, and showing evidence of an infection with a gas forming organism (a really bad sign). In the hours before she went to the OR and after starting the protocol, I was watching her closely. Her condition seemed to be stabilizing even prior to going to the OR. Although I was not at all convinced that Paul's protocol was as miraculous as his paper and the nurses on TV said it was, I was definitely intrigued; when she survived, that intrigue turned to genuine hope.

What happened next was transformative. A few days later, I administered the therapy to a newly admitted ICU patient with severe septic shock. He was a sixty-five-year-old man, seven days post–bone marrow transplant. Patients in the first seven days after transplant typically have no white cells to fight off infections and bacterial sepsis is a common complication. And boy was he septic. He was on high-dose intravenous vasopressor therapy, had altered mental status and labored breathing, and his kidneys had shut down. His wife was at his side, terrified that he was going to die.

So was I.

He looked terrible. I started him on what we later dubbed the HAT protocol (hydrocortisone, ascorbic acid, i.e., vitamin C, and thiamine) and what happened over the next few hours was something I had maybe seen once before amongst the thousands of patients I had treated for septic shock.

The nurses reported a rapid decrease in the need for vasopressors, an abrupt resumption of urine flow, a clearing of his mental status, and an easing of his breathing. I was thrilled by his progress, but that was nothing compared to what happened the next morning. He was the first patient I checked in on, and I was shocked to find the man sitting up in the armchair next to his bed, a tray of food next to him, eating breakfast, talking pleasantly with his wife. He was off all vasopressors and there was a full urine collection bag at the end of his catheter. The nurse informed me that he was being transferred back to the bone marrow transplant ward. Less than twenty-four hours from arrival in the ICU in severe septic shock?

I was exhilarated. *Holy crap. That stuff works!* I had never discharged a patient within twenty-four hours of neutropenic septic shock. Dr. Mark Juckett, the bone marrow transplant attending physician and a colleague I knew well, approached me. "What did you do to that guy?" he asked. "I thought for sure he would be on a ventilator and dialysis and he's actually going back upstairs?"

I blushed in almost embarrassment as I replied, "Mark, I'm telling you, I gave him high dose intravenous vitamin C and it turned him around!" The attending wasn't really sure what to say to that, so he just shrugged and mumbled "great" and continued on his patient rounds.

It was striking that an experienced bone marrow transplant attending had also noticed the sudden and unexpected physiologic reversal in such a short time. Suddenly, the word *miraculous* didn't seem so far-fetched.

Feeling as if I'd uncovered the key to the universe, I continued deploying Paul's HAT protocol and seeing dramatic clinical responses in extremely ill patients. I started a research study to collect data retrospectively in order to measure the outcomes of patients who had been treated with his protocol compared to patients who had not. I should note that within a couple of weeks, I started to see that in some patients, there was no response or minimal ones. I wasn't sure why and was a little concerned because I had become highly confident in the protocol's efficacy; cocky even. Every time I started the therapy, I had developed a habit of predicting to the nurse or my physician trainees what would happen next. Although I did have some concern over the occasional lack of response, the vast majority of patients receiving the protocol demonstrated dramatically positive alterations in their clinical trajectories.

I could not stop thinking about how absolutely life-altering this was. I had been treating septic shock for well over a decade and now had a treatment that was turning almost every patient around, successfully and quickly.

I finally decided to reply to Paul's email of almost eighteen months earlier. I told him how much I appreciated his congratulatory note on my editorial, and how my lack of reply was solely due to the circumstances surrounding my daughter's illness. I explained that I simply had not had the spirit or the emotional stability to write back to him at the time, but I was writing now because I wanted to talk to him about how IV vitamin C had completely transformed my practice and understanding of the treatment of septic shock.

His email reply to me was incredibly sweet and understanding; he gave me his phone number and said we should talk. I still remember that first conversation and exactly where I was and how we spent over two hours on the phone. He detailed what he was seeing in his ICU, how almost nobody required dialysis for acute kidney failure anymore and how the hospital nephrologists were noticing—and not necessarily in a good way. (A significant source of their income is from the reimbursement they get from providing acute dialysis.) He explained how his septic shock patients' average ICU stay was now *less than two days*. It was unreal. I was blown away by all of it, and thrilled that Paul's clinical experiences paralleled my own, although mine not as consistently. It would only be later that I would find out why this was.

When Paul shared with me data his CEO sent him, I actually got goosebumps. Apparently, someone at the largest data contractor for the Medicare and Medicaid programs, a company called Truven, had heard about his study (probably via those nurse interviews on TV!) and began tracking the mortality data of Medicare patients at his hospital over the next year. This was unsolicited data, submitted by an independent contractor. Paul had nothing to do with it—so zero conflict of interest—and it was astounding. The mortality rate of hospital sepsis patients had dropped from 22 percent to 6 percent in just over a year.

SNGH – Hospital Sepsis Mortality

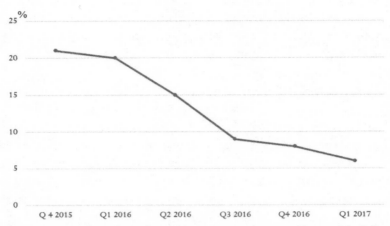

Figure 1. Truven Analytic data of mortality rates over time at Sentara Norfolk General Hospital.

I probably don't need to point this out, but I will: A six percent mortality rate for hospitalized sepsis patients is unprecedented. It also bears mentioning that Paul treated his first patient with the combination of IVC and corticosteroids in early 2016. After seeing the response in just *that one patient*, his protocol became standard practice in his ICU. Way back then, doctors could still use their powers of clinical observation to guide their practice—the standard method for medical progress for millennia, incidentally. Compare that to today's "wait for a [likely corrupt or fraudulent] Randomized Controlled Trial (RCT) to tell you what to do" and you might begin to understand the sad state of the system.

Doctors wrote to Paul from around the world. He collected more than 1,100 anecdotes at that time, almost all from emergency room or ICU doctors who had tried his protocol in patients and saw similarly dramatic clinical recoveries in severely ill patients, many of whom were often in multi-organ failure.

I was so consumed by these developments and my newfound alliance with Paul that it was almost all I could talk, think, or read about for months. I was still serving as the chief of the Critical Care Service and medical director of the main ICU at UW, and I began telling colleagues about my experiences and giving lectures. Eventually I put together a proposal to my entire ICU service of eighteen intensivists and nine fellows: We should, I insisted, start giving the therapy to everyone upon ICU admission. I tried to base the argument as simply as I could, citing the decades of studies showing severe vitamin C deficiency in almost every critically ill patient on arrival to the ICU (as the vitamin is rapidly consumed in critical illness). What I was thus proposing was merely a "repletion protocol." Not only was it patently safe, but I had also seen the dramatic results firsthand. I thought it was a slam dunk.

It turned out to be more of a brick shot. To say my superiors were not on board would be a gross understatement. In fact, one called my proposal unprofessional and just this side of quackery. This was, I'll remind you, an academic medical center populated by a cult who worships "evidence-based medicine" and conservatism and vehemently rejects any change to the status quo. Unsurprisingly, my suggestion to start repleting vitamin C in all septic shock patients was ignored by most (but not all) on the ICU service I was supposedly leading.

Despite the lack of uptake in the ICU, the ER docs downstairs were hearing about the HAT protocol from their residents who trained under me during their ICU rotations. I was invited to give what's called a grand rounds lecture to the emergency department so that the therapy could be started early, before ICU. Although we knew that early intervention was critical to efficacy, we had not yet determined how late was too late.

Later, I did a study which found that the impact of IVC on mortality was fundamentally dependent on exquisitely specific timing. During and after that time, numerous trials were conducted and published showing that IVC was ineffective in sepsis. All but one (still on a preprint server and thus not yet peer reviewed and published; hmm, I wonder why?) gave IVC far too late into severe sepsis. The preprint trial which gave IVC early was the only one that showed a massive mortality reduction.

My research found that for IVC to be effective in severe sepsis, it had to be given within twelve hours of presentation to a hospital. Further, we found that almost everyone survived if given IVC within six hours of arrival to the emergency department. Our data showed that if given after twelve hours there was no effect on survival (although other clinical impacts were still seen in those patients).

We had—and have—an intervention that could effectively obliterate the death rates of sepsis across the globe, yet only a few critical thinkers have paid attention to the data and actually use it. Just a month ago, on the front page of the *New England Journal of Medicine*, the top-rated medical journal in the world, a "rigorous" study of IVC managed to not start treatment until twenty-two hours after admission and included a significant number of patients who were already languishing in another ICU before transfer. Further, a cohort had surgical sepsis which is categorically different from medical sepsis in that patient outcomes are much more dependent on the surgeon's skills and the timing of the operation. So, guess what the study showed? That IVC *increased* mortality. I am not making this up. Basically, the *NEJM*—the veritable *Bible of Medicine*—concluded that intravenous vitamin C kills people.

They basically canceled IVC.

It's as infuriating as it is tragic, and as soon as we wake up from this Covid nightmare, Paul and I plan to resurrect the cause. I'm sure we'll rattle those rusty hospital cages again like we did with ivermectin when families of Covid patients hired attorneys like Ralph Lorigo to sue in court to be allowed access to the drug. The hospitals fought Ralph and those families to death. Oftentimes, literally.

The problem with asking the courts for help in treating a loved one with sepsis is that they'll never issue an order within six hours of the patient's arrival to the ER. Even in the absence of robust supporting data beyond our clinical experiences, we still believe that if IVC is given later, it can increase survival odds—but it would likely need to be administered at higher, even "massive" doses. I'd categorize it as can't hurt; will likely help.

Here's an analogy: You're on a booze cruise with your buddy when he falls overboard. He's drunk—and he's drowning. You search for a life preserver and discover that the "regulation" round one is missing, but there's a square-shaped one nearby that looks as if it'll do the trick. Some passengers are even saying they've known people who were saved with a square life preserver! If you do nothing, your friend is going to be fish food, possibly within minutes. Give me one good reason you *wouldn't* toss him the square one.

It boggles the mind. And yet this is what Paul and I are up against every single day. I believe our paths crossing was some sort of divine intervention because God knows I couldn't—and wouldn't want to—do this without him. What started with a spontaneous email and a long overdue follow-up has grown into a friendship, mentorship, and partnership that collectively grow stronger every day. Paul and I have supported and will continue to support each other intellectually and emotionally through our personal and professional Covid journeys and beyond.

For that, I am one lucky fecker.

CHAPTER THREE

The Long Road to Med School

Growing up, I wasn't one of those kids who had figured out his future career path by junior high. Far from it. Well-meaning adults would ask, "so, what do you want to *do* with your life?" The problem was, even well into college I was still erratic, unfocused, and hedonistic, with the typical undeveloped forebrain of a male young adult.

In other words, I was a hot mess.

Although I was smart enough to crush standardized exams without a worry, I had no idea who I was or what I wanted to be. A disciplined student I was not, but I did learn several important things in college: I hated business, sales, and a sole focus on making money. I loved literature but hadn't honed any writing skills of my own. I wasn't interested in politics or any field that was not directly or immediately impactful to another person. Eventually, I decided that I would either become a doctor like my father or a teacher like my mother. I thought those were meaningful pursuits and fit with what I enjoyed (math and science) and was good at (discussing, teaching, and learning with others).

Since I had a knack for numbers, particularly calculus, I ended up earning a degree in mathematics from the University of Colorado Boulder. Despite its widely heralded math department, Boulder—like seemingly every other US college—was a massive party school, and I was highly . . . social (ahem). Although I managed to graduate, I did it with steadily declining grades. I walked two years later than expected with a not-so-impressive 2.67 GPA.

I left college with no direction, no discipline, and no real goal. I was often unemployed and when I did manage to land random jobs like selling life insurance or doing data entry, I couldn't stick with them. I started to get depressed. I was still living at home, and insanely jealous of friends who had made it out of their parents' houses. I really wanted to become a responsible and productive member of society. At twenty-three, it was time to grow up.

I knew some restaurant guys who made a decent living, and while waiting tables was hardly saving the world, it would allow me to interact with people and hopefully impact their days—or at least their dining experiences. So I smooth-talked my way into the business, landing a job at Trattoria Diane, a high-end restaurant in Roslyn, New York. I had never waited tables before, and I had no idea that restaurant would change my life the way that it did.

The owner, John Durkin, would later become my best friend and mentor (and after that, the long-time mayor of Roslyn). Although he was eighteen years my senior, John had been just like me at my age, so he understood me. He's one of the most widely read people I've ever met and deeply intelligent, and we first connected on topics like music and literature. As we became closer, I started to seek out his guidance on a number of matters in my personal life that John had successfully navigated when he was my age.

John essentially mentored me in how to become a mature, responsible man and productive member of society. But when he initially hired me, I was lost. He actually came close to firing me several times in the first six months for showing up late, hung over, unfocused, or all three. But, little by little, I began withdrawing from the social scene I was devoting too much time to and instead started to focus on activities that would help me reach my now established goal of becoming a doctor.

I made a practice of staying home and reading a lot; I even started meditating. By changing my behaviors and activities, my life started to improve. I was given more and more responsibility in the restaurant and quickly began to earn enough money to rent a place of my own.

Eager to "erase" my undergraduate GPA, I went to grad school to study health administration, thinking that maybe excellent graduate school grades would help my chances of getting into medical school. I did that full time while also working in the restaurant full time. This was my first (but far from the last) period of overwork in my life. With the stress and intensity of two full-time pursuits involving a lot of commuting around NYC, I developed a severe tooth grinding habit along with massive dandruff. One night as I was going to bed, I passed a mirror wearing my dandruff medicine shower cap and bite guard. I looked at myself and thought, *what the hell are you doing to yourself, man?*

But I didn't care. I was happy in a novel way: I had discovered how satisfying it was to make positive contributions, at work and at school and at home, and to actually be valued for those contributions. I was motivated, focused, and committed in a way I had never been before. In my first year of grad school, I was hired by one of my professors to manage one of her research projects. I'm ashamed to admit that the project was CDC funded and focused on studying various financial and other incentives to physicians for *improving immunization coverage rates in the inner city.* I know. Or I should say, you don't know what you don't know.

I ended up getting my master's degree after earning straight As, something I hadn't done since the eleventh grade.

With that hard-earned piece of paper in hand, I started applying to medical schools in the US. Despite strong admission test scores, apparently nobody was impressed with my 2.67 undergrad GPA. I collected rejection letters the way a bikini-clad Instagram influencer racks up followers.

I ended up making an appointment with a college career advisor, looking for some practical advice on how to improve my chances of getting into medical school. He suggested seeking medical training overseas. It was a fantastic idea and something I hadn't considered. I ordered a booklet about offshore medical training opportunities and researched schools in places like Dublin, Israel, and Guadalajara that accepted American medical students. In the end, I chose St. George's University School of Medicine in Grenada, West Indies. It was a charming campus in a tropical paradise and pretty much the polar opposite of New York City. I'd been a life-long, obsessive windsurfer (later turned kitesurfer), and Grenada—with its crystal-clear Caribbean water—was my personal paradise. I windsurfed nearly every day and studied at night and couldn't have been more content.

I was twenty-eight when I started medical school in Grenada (nearly all of my classmates were twenty-two or twenty-three), and my experiences over the next four years could fill a book of their own. I completed rotations in St. Vincent, Barbados, and the UK, and appreciated learning medicine in such drastically different health systems. I graduated at thirty-two and landed an internal medicine residency spot at St. Luke's-Roosevelt on New York City's Upper West Side near Harlem, and then a fellowship training spot at Beth Israel Medical Center on the Lower East Side.

Looking back on my choice of specialty, I picked a doozy. When physician specialties are ranked according to an index of earnings versus quality of life, the top four are what intensivists (including me) derisively call the "ROAD scholars:" Radiology, Ophthalmology, Anesthesia, and Dermatology. These are highly desired and competitive because of the salaries they earn versus the stress and number of hours they typically require. In addition, they carry less burdensome "on call" responsibilities like covering hospital units and acutely ill patients off hours, overnight, and on weekends.

The "ROAD scholars" generally had the most manageable work hours while being the most highly paid. Conversely, at the bottom of the fifty specialties on this scale sit three: critical care doctors, family medicine doctors, and pediatricians. Apparently, we have the worst quality of life compared to our earnings. But to be fair, of the three at the bottom, ICU doctors make the most, so don't bring out the violins. In fact, ICU doctors, compared to most specialties, do very well in terms of income, but when you factor in the hours and stress, it drags us to the bottom of the most desirable specialties. I know this from coming home after a "normal" day in the ICU, unable to sleep while agonizing over what I may have missed in a patient dying under my care. It was insanely stressful, especially in my early career when it's normal to find yourself unsure about a lot of things. I would fall asleep and wake up thinking of whichever patient I was most worried about. On the train into work, I would research my hypotheses for each patient's illness and trajectories, which always precluded me from working on a paper or reading the news.

Although my specialty may seem like a poor choice, I loved ICU medicine. During my general medicine training, I was in awe of the ICU docs; I thought they were the "baddest of the badasses." I was intimidated by and in admiration of their broad knowledge base and diversity of clinical skills. I also admired that staff from every area of the hospital, from labor and delivery to rehab, invariably called for an ICU doctor when a patient suddenly took a turn for the worse. The ICU docs were experts at the widest variety of illnesses, especially in their most severe forms. They would stabilize and treat the sickest of the sick. They were heroes.

I wanted to be like the intensivists who could so calmly and expertly navigate those stressful clinical situations. I admit I was quite intimidated when I made that decision; I wasn't sure I had what it took. My wife, herself an intensivist by that time, offered her wholehearted support and reassurance and encouraged me to follow in her footsteps.

Quick side note: One of my greatest achievements in life was asking my wife Amy out for the very first time. I made this embarrassingly bold move as a fourth-year medical student, which means I was wearing a *short white coat*. Worse, it was a neurology rotation, and so I had a dorky reflex hammer sticking out of one of the pockets. At the time, my wife was a senior resident at the end of her training and thus had earned the right to wear the magnificent and coveted *long* white coat. If you aren't in medicine, you may not fully grasp the chutzpah it took for a medical student *in a short white coat* to ask out a senior resident physician. Like my best man, John, the mayor of Roslyn from my restaurant days, said at our wedding, "I always wondered why Pierre was so nervous to ask Amy out, but I finally got it. In my business, it would be like the dishwasher asking out the chef!" Exactly.

On top of the incompatible jacket issue, Amy was—and is—intimidatingly beautiful. It was a bold move in every way.

I was three years behind Amy in training and she was one of the best doctors I knew. She would be up all night managing patients by phone at four different New York City hospitals, only to get up later and put in a full day in the ICU. She was a rockstar—and I wanted to be like her. I didn't give a passing thought to how challenging it would be raising a family with a wife who was also an ICU doctor in poorly resourced and uberexpensive New York City.

So I followed in my wife's footsteps, obtaining a fellowship training spot at Beth Israel Medical Center in 2005. Three years later, I had completed my training with three and a half board certifications under my belt: Internal Medicine, Pulmonary Medicine, and Critical Care. The half refers to the fact I was granted what is known as Testamur status in adult echocardiography from the National Board of Echocardiography, the highest ranking a non-cardiologist can achieve and one of the most challenging accomplishments of my career.

At this point I was thirty-eight years old, living in NYC, married with two young daughters, $150 in my bank account, six-figure, double student loan debt, and not a penny in savings. It was not exactly what they call *killing the game*.

It was time for this former party boy to get a job. There were basically three categories of physician career paths to choose from: pure clinicians, who devote their time exclusively to patient care; clinician-educators, who care for patients while also researching and teaching that care to physicians in training; and physician-scientists, who split their time between writing grant proposals for research funding, conducting research experiments in their labs (why they are called "lab rats" by us clinicians) and publishing papers. As a result, only a minority of their time is spent doing patient care and teaching. This last group, who often have PhDs in a basic science discipline beyond their medical degrees, are only found within large, academic research institutions and are by far the system's darlings given that when they win a research grant from the NIH or Pharma, it brings a lot of money to the institution. Because of this, they are generally the only ones who rise to leadership positions within those institutions.

One of the core responsibilities of a physician, enshrined in the Hippocratic Oath, is to *add knowledge to the practice of medicine*. I always took that part of the oath very seriously and essentially committed my career to it, not out of some moral or ethical imperative, but because I loved teaching, research, and clinical innovation. Thus, I decided I wanted to become a clinician-educator. Clinicians interested in teaching the skill do so via medicine's historic apprenticeship model, which works as follows: Our students and residents or fellows see our patients first and then present to us the information they have gathered along with a proposed care plan. Then they observe us while we evaluate and interact with patients and decide on the ultimate treatment approach. It's extremely hands-on and how I trained as well, under the most masterful of mentors. By the time I left academics, I had mentored apprentices for fifteen years.

One of the proudest achievements in my career as a clinician-educator was becoming one of the youngest program directors of a pulmonary and critical care fellowship training program in the country. Although my official title was associate program director (I was only in my fourth year as an attending and the position required a minimum of five years of experience), for all intents and purposes, I was the PD. Another was winning departmental teaching awards at every academic medical center I worked for. However, what I am most proud of is helping pioneer two different, then-novel fields of my specialty. The first was on the study and application of therapeutic hypothermia in post–cardiac arrest patients, and the other was creating and teaching a novel diagnostic approach to critical illness using what is known as point-of-care ultrasonography.

My interest in therapeutic hypothermia developed early on and stemmed from a case of a patient who had suffered cardiac arrest on one of the hospital wards. We successfully resuscitated him and brought him back to the ICU but, as is typical, he remained in a dense coma with little detectable brain function. The downtime of the arrest was almost thirty minutes, which is not good prognostically. My mentor, Paul Mayo (whom I will subsequently refer to as Dr. Mayo so as not to confuse him with Paul Marik), told me that we should "cool" him—meaning drop his body temperature to cool his brain, which for simplicity's sake is akin to icing an injured knee in terms of reducing the damaging inflammatory response. Dr. Mayo had recently learned that this was something European doctors were beginning to do with comatose post-arrest patients with success. So we cooled the patient, and three days later he regained full consciousness. Moved by this clinical experience, I dove in and started researching everything that was known about the therapy.

As a result of that obsession, I created the first therapeutic hypothermia protocol in a NYC hospital. I quickly became a regional and then national expert in the therapy and started lecturing on the topic. I conducted research studies trying to find which patients benefited most or not at all, and on whether the speed or depth of cooling mattered. I was one of the panel members that helped develop NYC's "Project Hypothermia," where, when we started, my hospital was the only one of the forty-six area hospitals that had a hypothermia protocol. Within a few years such protocols had been implemented at every hospital in the city and all ambulances and paramedics were equipped with and trained in the cooling of post–cardiac arrest patients.

As more and more studies were done, my interests and expertise in therapeutic hypothermia waned as I discovered that it was largely beneficial only in out-of-hospital patients with an underlying cause of heart disease, rather than in my general post-arrest ICU patients. Then studies showed that cooling such patients was less important than simply preventing rises in temperature, so now cooling approaches are more reactive than proactive and essentially focus on maintaining "normothermia," or normal core body temperature. However, it was a great early professional experience.

My next major career milestone was when I became one of the world experts in pioneering the development and teaching of what's known as point-of-care ultrasonography (POCUS). POCUS is the use of ultrasound by physicians at the bedside—the literal point-of-care—to image internal organs in real time and thus make rapid, accurate, often life-saving diagnoses in patients. POCUS has the greatest impact in the gravely ill, as the timeliness of both diagnosis and treatment in those patients are critical to their survival. An accurate diagnosis arrived at in a delayed fashion makes it too late to impact their clinical trajectory. A wrong diagnosis and treatment early on could seal their fate.

Prior to the advent of POCUS, ultrasonography involved getting an ultrasonographer on site, waiting for his or her findings to be sent to a radiologist, and then waiting for said radiologist to interpret and finally (rarely in due time) write up and send a report. This painstaking process could take hours, all while the treating physician would be sweating it out at the bedside, watching patients fight for their lives.

Understand this: Physical exam findings—palpation and auscultation using hands and the stethoscope—are extremely limited in determining the proximate cause of a life-threatening illness. As a doctor of the dying, you need to know real time functioning of the critical, life-sustaining organs: the heart, lung, liver, kidneys, intestines, and large blood vessels. Blood tests may reveal an abnormality but not necessarily its cause. Physical examination will give you clues but rarely a definitive answer. With an ultrasound probe in your hand, you can image the entire contours and function of those organs and the information gained can lead you to initiate targeted therapies to reverse the respective organ failure detected.

POCUS was absolutely magical in the type, amount, and speed of information that it provided. I discovered this "magic" early in my career and could not believe that the entire world of ICU doctors was not utilizing this technology. I had many colleagues who, after relying on traditional methods of observation and intuition and reasoning, all of which were highly variable, would sign out their service to me on a Sunday night (ICU docs typically run the ICU for a week at a time and would switch on Monday mornings). I would then come in on Monday and discover abnormalities they had not been aware of. It was terrifying, knowing that the average doctor was relying on imperfect information while I had access to highly accurate, real-time, and often treatment-changing information.

Over the next fifteen years, I would devote much of my career to teaching doctors across the world how to use ultrasound in the care of critically ill patients. But when I began, I knew very little, obviously. I had to make myself expert in the skills of image acquisition and interpretation. Essentially, I had to master the jobs of both an ultrasonographer and a radiologist. In the US and many other countries, these skills are never taught to non-radiologists, so I was forced to learn largely on my own. An autodidact, if you will. Aside from my mentor Dr. Mayo, there were few ICU medicine practitioners who knew how to create ultrasound images in real time and interpret them accurately.

Dr. Mayo taught me everything he knew about ultrasound before he left to take a position at a different hospital, first showing me the basics of acquiring and interpreting images of normal and diseased organs. Ironically, he himself learned ultrasound from a German intern. In Germany, ultrasound image acquisition and interpretation were a standard part of a doctor's medical training, starting in medical school. Although the average level of skill among young doctors there was quite low, the concept was highly valued. This was not the case yet in the US.

I don't think anyone outside of medicine can understand the uniqueness of what Dr. Mayo did with ultrasound. After all, he was a professor and the director of an ICU in a major metropolitan center when *an intern* trained in Germany informed him of the benefits of learning to use ultrasound. You could determine if a patient's lungs were dry or wet, if their gallbladder was inflamed, or if their heart was functioning and if not, which part of their heart was the culprit, or if they had blood clots in their veins. The amount of detailed, specific information you could acquire within seconds to minutes was incredible. And the specificity of the information blew away the frequent inaccuracy and non-specificity of chest and abdomen X-rays.

To his historic credit, Dr. Mayo listened to this intern. He started an "ultrasound club" where he would take an afternoon each week to practice image acquisition with the intern and any others who were interested. After learning what he could from that intern, Dr. Mayo reached out to his cardiology colleagues and began going down to the Echo lab reading room to interpret cardiology exams with the experts. He even got them to teach him trans-esophageal cardiac echo (which was even more complex and invasive).

In 2000, Dr. Mayo and his senior fellow at the time, Dr. Adolfo Kaplan, became two of the nation's first non-cardiologists to pass the National Board of Echocardiography exam, one of the hardest exams among all specialty exams with a 65 percent passing rate even amongst cardiologists. I later joined that elite group when I too passed the exam *by one question* in 2008.

Dr. Mayo then became colleagues with the trailblazers in the field, which were the French—who interestingly refer to ICU doctors as "re-animators." The pioneer and world-expert leader of POCUS among the French intensivists was Dr. Daniel Lichtenstein, a savant in the truest sense of the word. Dr. Mayo became close friends and colleagues with Daniel. I read his seminal textbook maybe six times, ingesting every word. One early honor in my career was when, as a fellow still in training, I gave a lecture at a conference right before his keynote lecture.

There's nothing in the world like being faced with a patient dying of multi or even single organ failure and *not knowing why.* Imagine you are the most senior doctor in the ICU and a team of trainees, nurses, and consulting physicians (not to mention families) are looking to you to diagnose and guide an effective treatment plan . . . and you don't know what is wrong.

Unfortunately, many doctors in such situations simply start barking orders, as if the appearance of doing something is somehow going to fix anything. They embody the adage of "don't just stand there, do something." No one asks the doctor why they order this test or that medication; they assume there are valid, logical reasons.

As I gained experience in critical care, when I was in those situations, I started doing the opposite. I became quiet and allowed my mind to race through the numerous diagnostic possibilities until I arrived at what I thought the best approach was. Yes, sometimes it included barking orders for testing/labs but it was mostly about absorbing the available information, asking for further details based on history, and then allowing intuition and pattern recognition to guide my next actions. Many times, the answer/diagnosis would come to me in this way only to be later validated by testing or the response to my treatment choice. What is fascinating about these situations is that they are unique to critical care, where we are often immersed in medical emergencies defined by chaos, urgency, and the need to act on incomplete and sometimes unknowable information.

After I became an expert in POCUS, these situations were greatly reduced in complexity and stress because often and within minutes, I would know exactly what was wrong and could turn patients around. I could quickly identify (or conversely and more importantly, rule out) a blown right or left ventricle, a massive pulmonary embolus, or excessive fluid in the lungs, among many other critical diagnoses. It was profound, not only to me and my trainees, but most importantly, to the welfare of the patient.

When I finished training, although I was not on a par with Paul Mayo, I was recognized nationally as a leading expert. So he and I, along with the amazing Seth Koenig, Mangala Narasimhan, and Robert Arntfield, learned and taught and published together for years. We put together the first local, then regional, then national, then international courses in critical care ultrasonography. We traveled the country and world for years teaching, researching, and publishing. We created and administered the first certification exams. In collaboration with the Europeans, we—not some bureaucrat or self-designed health care leader—established the "standard of care" and "consensus opinions." It was exhausting but exhilarating, because it was new, and because we were setting the standard for the practice of POCUS.

One of my proudest achievements around POCUS was when I was invited to become the senior editor and coauthor of the now best-selling textbook in the field. Currently in its second edition and having been translated into seven languages, *Point of Care Ultrasound* won the British

Medical Association President's Choice Award for best textbook the year it came out. I wish I could take credit for its impact but Dr. Nilam Soni, a similarly obsessed POCUS expert, deserves all the recognition. He is one of the most amazing of the many phenomenal doctors I have had the honor of working with. Yes, it's a brag; but I am genuinely humbled.

Pioneering in medicine is NOT easy. My early years learning and practicing were also spent fighting radiologists and cardiologists who wanted to protect their turf and continue using (and billing for) ultrasound to look at the heart and other organs. They said we could never become sufficiently competent and that we would hurt patients instead by making inaccurate diagnoses. I said fine, you can come and do your comprehensive heart exam at three in the morning while someone is deteriorating in front of me. Not one of them ever took me up on that invitation.

Their obstructionism was driven largely, in my mind, by fear that we would order less "official" and also less billable ultrasound exams. It took us a long time, but we eventually proved them wrong in regard to making errors. I will say we certainly may have missed subtleties on those exams, but subtle findings are almost never immediately life threatening. On the other hand, their fears of losing reimbursements were very real as I definitely ordered a lot fewer formal ultrasound exams after I developed the imaging skills myself.

I also engaged in some epic fights with hospital administration, trying to get them to open the pocketbook and buy ultrasound machines for use in the ICU. But by arguing that it was critical for patient safety—which would ultimately reduce procedural catastrophes and lawsuits, every hospital's greatest fear—eventually we succeeded in equipping every ICU in the country with an ultrasound machine as the standard of care.

These early career exploits developing and teaching innovative approaches to medicine around the country, combined with an overwhelmingly busy outpatient and ICU practice, eventually led to full-blown burnout. Doing all the above while also commuting three round-trip hours a day led me to a point where I allowed myself to be recruited by the University of Wisconsin.

A life-long New Yorker moving his family to *Wisconsin?* What the hell was I thinking? (In my defense, my wife is from there.) I didn't know it at the time, but the academic career that I embarked on at thirty-eight years old would end there—and mark the beginning of the tumultuous and exciting next chapter of my life.

CHAPTER FOUR

The Legacy of Paul Marik

One of the things that surprised me throughout our later Covid journey was how the prevention and treatment guidance that Paul Marik, and later the FLCCC, disseminated was attacked and then dismissed with such animosity, despite our cumulative career achievements and credibility. None of us enjoyed the derision, but the attacks were most surprising in regard to Paul. He literally is (and was) the most published practicing critical care doctor in the history of our specialty. In addition to being globally renowned for his revolutionary work in sepsis treatment, his knowledge base, academic background, and contributions to our specialty are unparalleled.

Paul has told me he is most proud of his work in teaching a global generation of critical care doctors that measuring central venous pressure (CVP) to estimate the fluid needs of a patient was useless outside of a very narrow set of circumstances. It's important to understand that CVP was used for decades in ICU patients who were in states of shock characterized by dangerously low blood pressure. It was the standard of care in ICUs. Paul did a deep dive into the published literature and wrote pretty much the most impactful paper ever called, "Does Central Venous Pressure Predict Fluid Responsiveness? A Systematic Review of the Literature and the Tale of Seven Mares." The paper's most memorable sentence was this: "The only study we could find demonstrating the utility of CVP in predicting volume status was performed in seven standing, awake mares undergoing controlled hemorrhage."

Mares. *As in horses.*

His paper triggered fierce debate in critical care circles that lasted for years. Reversing established orthodoxy in medicine is nearly impossible. But Paul single-handedly pulled it off with his papers and lectures, helped by a lot of folks like me who followed his work closely. I would argue that today, the obsession with using CVP to guide fluid resuscitation has largely (but not completely) been abandoned.

That was all Paul.

When it came to how the specialty of critical care treated the condition of sepsis, Paul was again a beast at demolishing the prevailing "wisdom" at national conferences. He was so good, his take on the data so expert and compelling, that his lectures were always packed. I'm talking standing-room-only packed. For a medical lecture.

The most debated aspect of sepsis treatment twenty years ago was called "early goal directed therapy" (EGDT) which, without going into it too deeply, dictated numerous interventions which made little sense in terms of necessity. Further, what the approach led to, Paul found, was a harmful excess of intravenous fluid administration.

Even as a fellow, I knew the approach was not physiologically sensible. My mentors knew it, Paul knew it, yet EGDT was widely adopted across the country and around the world. The protocol was based on a single center study whose principal investigator Manny Rivers *held the patent on one of the pieces of equipment necessary to execute the protocol,* a fact unknown by most at the time. Further, information later emerged showing that the data from the original trial had been manipulated, information that was leaked by a whistleblower who was a fellow of Rivers at the time. The hospital threatened the fellow with the ending of his career if he were to continue to speak publicly about it. They even apparently threatened to kill his kids.

Sound frighteningly familiar?

Another aspect of the US sepsis resuscitation guidelines that Paul was absolutely genius in debunking was the decision to target a reduction in lactic acid as the immediate goal. While this ultimately became the "standard of care," Paul's research revealed that targeting lactate was the result of a gross misunderstanding of lactic acid physiology. It was another of the most masterful papers I have ever read.

Just like with his teachings on CVP, again you had one man arguing against an entire generation of doctors who believed that reducing lactate was important in the general septic patient. This time, however, despite all of Paul's research and lectures, today doctors across the country are mandated by government payors to repeatedly check and respond to lactates in septic patients. Otherwise, the hospital receives poor quality scores which can affect their reimbursement. It is yet another example of an orthodoxy based largely on fiction.

Although he was unsuccessful in changing this now-ironclad lactic acid orthodoxy, Paul's efforts led to widespread admiration and reverence for his comprehensive knowledge and scientific acumen.

That was probably nice while it lasted.

Alas, eventually Covid would come along. Whereas Paul has previously bucked pervasive misunderstandings and applications of physiologic beliefs, now he was trying to bust the deadly, erroneous myth that a simple, inexpensive, readily available medication was ineffective in treating Covid. That effort ended his career. Instead of fighting widespread yet ignorant knowledge of physiology, he was now up against the vested interests of Big Pharma.

They say courage is grace under pressure. Paul Marik is a brave and powerful force, and one I believe history will recall with profound respect and admiration.

CHAPTER FIVE

Covid Hits the US

Soon after returning to the US from the disappointment and drama of the conference in Belfast, I started hearing of a new and mysterious viral illness epidemic in China. People were filling hospitals and dying of lung injuries. Memories of all the hysteria and reports of SARS in 2003, H1N1 in 2009, and MERS in 2012 came flooding back. The initial intrigue I felt rapidly turned to concern as I was bombarded with images of overflowing ICUs with all the providers in what looked like hazmat suits, the likes of which I had never seen.

At first the craziness seemed both rare and far removed, like so many tragic things constantly filling the newspapers. It would have been almost easy to brush aside if I weren't consumed with all things medical in general and lungs in particular.

The stories kept getting worse, with Chinese cities going on lockdown, military spilling into the streets, and hospitals overflowing. Next it was Lombardy, Italy, making headlines, complete with images of hastily constructed outdoor tents full of patients on ventilators. I had never heard of ICUs filling up with such speed. Further, I was hearing concerning reports of therapies not working and leading to extraordinarily high mortalities once patients landed on a ventilator in the ICU. This was totally discordant with the decreasing mortalities we have proudly achieved over the last twenty years in ICU care and with advances in ventilator techniques. The world started locking down, effectively pretending that a highly transmissible virus could ever be contained in such a crude and ineffectual way.

At the time, there was supposedly only a Covid case or two in the US, so to many this was still "their problem." We had a hospital ICU leadership meeting to discuss the virus in late February, and my boss reassured all of us that it "wouldn't be coming here." I was unsure. With China and now Italy and international travel so widespread, I was pretty sure there was no way we could avoid the calamity that was coming.

Soon, Seattle's hospitals and ICUs began to fill. Several of my colleagues at UW had been trained in Seattle and knew doctors working there. Word from the front lines was that it was absolute chaos, requiring constant out of the box thinking to solve shortages in beds, doctors, nurses, space, and equipment while trying to protect patients and providers against an invisible invader. No one knew if the virus was being transmitted by large droplets or aerosolized droplets or from touching surfaces. My sense was that this thing was airborne, meaning you could get it from simply breathing the exhaled air of another person in the same room. It was terrifying to consider.

Knowing what I know now, these scenes of chaos were not universal at all. Many towns, cities, hospitals, and health systems handled that initial wave without incident. Still, we were operating on red alert at all times, traumatized by the incessant fearmongering across every form of media.

And so began my Covid journey. It started with the slow realization that the world was being overwhelmed with a severe pulmonary and critical care disease. I was a *pulmonary and critical care expert* at the peak of my career at one of the major academic medical centers in the country.

Game on.

Paul and I started talking every day, sharing the things we were learning from the news, journals, and preprint servers as well as from doctors around the country and the world who were starting to face off against the disease. I knew what was coming and that I would be on the front lines and that I had to figure this disease out, and quickly.

I will admit that one of my first acts was to design a protocol for myself to give to my wife so that she could ensure I was aggressively treated if I became infected. (I'd know how to treat *her* if the situation were reversed.) Newly overweight in the aftermath of the traumatizing years I had spent fighting the medical system for my daughter, I was scared. But I also knew it was critical that as an ICU expert, I had to do everything I could to figure out how to treat this disease. My initial protocol largely consisted of corticosteroids, a cytokine blocker, an anticoagulant, and intravenous vitamin C, all of which have now proven efficacy in treating Covid. It was one of Lucky Pierre's earliest wins.

New York City was next on Covid's hit list. Hospitals were blowing up with cases in my hometown and training ground. Dozens of friends and colleagues whom I had trained with, trained under, or trained myself were manning ICUs in both frontline and leadership positions. The stories were damning, dramatic, and getting worse.

The news was a nonstop nightmare. Jobs were being lost, the stock market was this-close to crashing, and Covid cases were skyrocketing. I was scared for the whole country. I'd read that a third of Americans do not have $400 available to cover an emergency. (That number would explode to 67 percent in 2022. Thanks, Covid!) How were people going to be able to miss weeks of work and still pay for food and rent? What would happen to every aspect of our health while trapped in our houses and apartments or worse yet, left on the streets?

My Wisconsin health care system and hospital started pulling me into a myriad of disaster planning and institutional response efforts, and all of the world's other emergencies faded into the background. Each day, each meeting, revealed yet another hole in our plan to manage the inevitable next surge. While I dealt with issues of ventilator, specialist, and gown-glove-sanitizer capacity, the epidemiologists and data scientists down the hall were trying to model the size and timing of the wave of patients predicted to hit our system.

It was reassuring that only one of the five models depicted a true Armageddon; the other four ranged from severe chaos to likely manageable. But the data, predictions, and ground realities were changing by the hour. The intensity of activity in my immediate surroundings consumed me. I was forced to put my family on the back burner, a problem that continues to this day, despite numerous attempts to make my life more manageable. I keep fantasizing (and promising them) that after I finish this damn book, they'll have me back.

A crazy, corruption-fighting physician can dream.

The only thing I could seem to find time for outside of work was talking to my colleagues on the front lines in New York ICUs. The war metaphor, although overused, appeared apt. Back in NY, my beloved mentor Dr. Paul Mayo got Covid early. We all checked on him daily, anxious when we learned he had developed shortness of breath or his oxygen had started to dip. Mercifully, he made a full recovery.

Dr. Mayo was one of the lucky ones. A small but nevertheless disturbing number of doctors and nurses were starting to die on the battlefield. It seemed like Russian roulette, with some fortunate folks getting blanks: a fever, mild cough, body aches, and a few days break from work. Others got the bullet: shortness of breath, chest pain, plummeting oxygen levels, sedatives, a ventilator, and sometimes death, all amid staff, medicine, and equipment shortages.

It didn't seem like things could continue to get worse, but they did. When he returned to work, Dr. Mayo described his ICU as rows of patients on ventilators in a large, open ward that had been quickly converted into a makeshift ICU. He told me that the majority were dark-skinned and that most were overweight or obese. All were on maximal support. Very few were coming off, and few were dying quickly. Most were stuck in ventilator purgatory.

These Covid patients presented a new trajectory of illness. Typically, and within a few days, severely ill, ventilated patients will either start to show small, steady improvements or experience a slow, steady decline to eventual death. One way or another, they will "declare themselves" within days to a week. Here, patients were deteriorating to the point of requiring maximal ventilatory support and then staying that way, without showing signs of recovery or further deterioration, often for weeks. It was bizarre.

Prior to Covid, patients in respiratory failure in my ICU generally were on ventilators for two to four days, a metric known as ventilator length-of-stay, or "vent LOS." Our ICU at UW prided itself on having a consistently low vent LOS compared to national metrics.

Contrast that with the fact that in Dr. Mayo's health system's two main hospitals in NYC, they had intubated their first Covid patient fifteen days ago and their tenth patient ten days ago, and *not a single one had been removed from the ventilator*. In the previous two weeks, they had accumulated seventy-one patients on ventilators, and none had been removed.

We were terrified. I led teams in my ICU figuring out how to set up beds so that all the IV pumps and tubing and even the ventilator control panel and monitor could be located *outside* the room, cutting down on the need for entering and exiting Covid rooms and the risks and resources each bedside visit entailed. At the time, Covid patients were all in isolation, which meant "donning and doffing" personal protective equipment (PPE) including gown, gloves, mask, and shield on every entrance and exit. There was so much fear of contamination and transmission that we adopted a complicated, multi-step, sequenced procedure of removing our PPE. It was both nerve-racking and exhausting.

I worked alongside pregnant colleagues; single moms and sole providers; the immunosuppressed; folks at the ends of their careers. All were aware of—and frightened by—the risks, but every single one carried on. The country was calling us "health care heroes." In New York City, every night at 7:00 p.m., all the fire trucks and police cars blazed their sirens, the cabs and cars blew their horns, and people cheered as the trucks and ambulances went by. It was a true sense of community, connection, and appreciation. It's impossible to fathom that that same system is the one that later became complicit in a global vaccination campaign that caused a humanitarian catastrophe whose implications we're only just beginning to see and understand. Today, almost everyone I meet (most of them unvaccinated and "awake") is terrified of going to a doctor or hospital. It's heartbreaking, actually.

I had never seen such insanely rapid changes to policies and practices. Noncritical tests all but vanished. Orders for chest X-rays, EKGs, physical exams, and blood tests plummeted. In my mind, this was hardly a bad thing. I was always a minimalist physician, ordering the fewest tests of anyone on my service as I had long been bothered by the tendency of US doctors to order unnecessary tests and procedures. Suddenly, the ordering of tests became right sized, used only for truly indicated reasons. It was magical and I loved it. I knew that, paradoxically, this practice delivered not only better care but was also a better use of resources.

One of our surgeons, Dr. Ben Zarzaur, had worked for ten years directing disaster teams at FEMA, where his sole mission was to safely extract people from rubble after a disaster. He felt we needed to reform our entire command structure and develop a surge response plan. In my experience, academic teaching hospitals are some of the most complex, fragmented, bureaucratic, and hierarchical institutions in the world, but suddenly those in power unleashed us frontline and clinical leadership with the autonomy to synchronize all of our schedules and teams and create an insanely complex master schedule across multiple departments and divisions. The idea was that all providers be interchangeable to be able to "plug and play" into any slot in the event of a physician falling ill. Further, we had reserves of non-ICU fellows, other specialty attendings, interns, residents, and surgeons who could fill frontline roles on the multiple critical care teams we were constructing for the surge. It was incredibly intricate and, it seemed, our only option.

One night, a former fellow who had recently graduated and was now a lung/ICU specialist in Detroit called me.

"I'm getting killed here," he said. "I need your help with these Covid patients." Apparently, they had fifty patients in his hospital on ventilators in acute respiratory failure. It was an astronomical number. He had tried numerous ventilator maneuvers, trials of paralytics, proning, and the use of sophisticated ventilator modes. Nothing was working. All of the patients' blood oxygen levels were still low and they were running out of pain and sedative infusions as well as critical antibiotics.

Even more worrisome was the fact that they had just used their last ventilator and had to take two patients to the operating room to "split" a ventilator so that it could support them both at the same time. This is a complex endeavor fraught with potential complications. Although I believed I could do it reliably because I had been studying various techniques with my colleagues in the event we came upon the same situation, it's a dicey proposition intended for extreme emergencies, like in the 2017 mass shooting in Las Vegas where a clever ER doc temporarily split two ventilators to support four critically ill patients.

The Detroit ICU doc and I spent a long time reviewing why and how the therapies he was trying had failed. I told him to stay calm and reminded him that blood oxygen saturations in the 80s (or even far less) in the setting of good cardiac function are well tolerated. Then I suggested treating these patients with . . . wait for it . . . corticosteroids and intravenous vitamin C, both therapies with large evidence bases of benefit in severe acute lung failure.

However, the infectious disease (ID) doctors at his hospital had been central in creating their own "institutional protocol." This had been happening since the beginning of Covid and it irritated the hell out of me. It seemed that at every hospital in the country, committees had formed within clinical leadership, and although some were recommending active treatment, the majority were engaging in supportive care only. Clown World had officially been built.

One thing that ID doctors are afraid of is corticosteroids. This is because although they act as anti-inflammatory drugs, they also have immunosuppressant qualities as they (slightly) raise the risk of superimposed infections. As an intensivist, I knew that despite these risks, they had an overall net benefit in many situations. Plus, out of all the specialists I worked with as a doctor running ICUs, the ones who frustrated me the most (with rare exceptions of course) were ID docs. They sent ridiculous batteries of tests and used massive amounts of anti-infective agents on my patients. It's a more-is-more approach to doctoring that I have always despised.

Naturally, the ID docs in Detroit were against the use of corticosteroids. I lost it.

"Then tell the ID docs to *get the hell out of the ICU*," I roared. He was the doctor in charge, not them. They were consultants, they could make recommendations, but the *Capo di tutti capi* of the patient is the ICU attending. He was new in his career and terrified to make waves, but my furious words made an impact on him, and he began treating the patients with corticosteroids. He would later call me to thank me, informing me that after he started the corticosteroids, his patients began coming off ventilators. Other ICU docs started to follow suit, and in his words, he was now viewed locally as a hero.

Lucky Pierre strikes again.

Another reason I was so livid at the time is that I already understood that Covid-19 in the ICU was *not* an infectious disease. Research had shown that in most patients, six days after first symptoms, no live virus could be cultured. Thus, the later hospital phase was largely marked by a massive inflammatory response to the virus or spike or mRNA debris. This is why corticosteroids are so effective in treating the illness.

Fast forward to late March 2020. At the time, I was drowning. Not in patients but in hospital preparations as well as another family health crisis. I was on a sort of half-leave as I had ceded my position to a colleague due to my second daughter falling ill with the same debilitating neurological syndrome (PANS) as her sister had. Even though the position I held was supposedly half time, it became so massive that it required four full-time people to plan and execute all the initiatives we were creating. I was working constantly and being barraged with hundreds of emails a day. Further, I was waiting to fight the fight of my life, awash in a mix of emotions from fear and dread to excitement and inspiration.

But my home was in crisis. We again had to immunosuppress my ill daughter to treat her condition and Amy was extremely nervous every time I came home from work. Despite the fact I was following a "protocol" of taking off all my scrubs in the garage, putting them in a bag, walking in the house in my boxer shorts, putting everything in the wash, sanitizing every inch of exposed skin and finally then getting dressed in clean clothes, I was still a risk. No hugs or kisses for anyone (instead I would kiss my daughter's back). We wore masks all the time when in the same room. Those were strange and scary times for sure, but this time at least, we knew what we were dealing with and how to treat it.

With my first cycle running the ICU coming up, Amy could no longer bear the idea of me treating Covid patients and coming home with it. She didn't even want me living in the house, so she asked me to move into a hotel.

A friend from Colorado texted me around this time, asking if I would share what I was seeing in my practice. I began the process of documenting what was happening in the ICU and emailing her each night. This is copied directly from those notes:

> Our surge here in Madison is predicted to hit in five to ten days and batter us
> for weeks. Milwaukee is starting to surge. Our blessing is that we have had
> a lot of time to prepare (relative term but in this crisis a week is a [expletive]

luxury) and so I don't think it will be like it is with my friends in New York City where it came on so fast all they could do was rapidly swell and strain their normal day to day operational resources and structure in a haphazard, emergent fashion.

But it's starting here. We already have two Covid patients in the ICU, several on the wards are not doing so well, and the first emergent intubation onto a ventilator occurred last night. I'm just glad we have a fleshed-out command system and structure and are rapidly preparing lines of reserves and back benchers to fill critical positions if we frontline pulmonary and critical care guys start getting sick and sitting out. It's estimated that 30–40% of normal workforces will be out sick during the surge with some (or even many) dying.

Remember, this was when we were not only awash in media fear porn, but the testing was sparse, and we had no idea of the true case fatality rate. During those early weeks and months, many of us ICU docs were on massive WhatsApp groups, sharing papers and insights and approaches and experiences. For a while, the messaging app served as a potent, real-time medical textbook.

My palliative care colleagues approached us lung doctors asking us for lists of our chronic, advanced lung disease patients, the majority of whom would not benefit from a ventilator if they got Covid and deteriorated to advanced acute respiratory failure. It was an attempt to preserve resources for those most likely to benefit (and avoid use in those with little chance of survival). They wanted to reach out to have advance care planning discussions with these patients, and to prepare to deploy hospice staff if that became necessary. This was sensible rationing, by the way, unlike the horror show of what later happened in the UK whereby, in the spring of 2020, their health system issued policies essentially making hospitals or ventilators unavailable to their entire population of care home residents, causing massive increases in death from both Covid and non-Covid.

It was around this time that I received this email from my Chair of Medicine, which rattled me to the core:

It is with great sadness that I announce the death of Dr. John Murray this morning from ARDS [acute respiratory distress syndrome] due to Covid19. As you know, Dr. Murray was one of the true giants in pulmonary and critical care medicine and made innumerable contributions to our field, nicely outlined in the attached note written by Dr. Courtney Broaddus. <u>It is both fitting and ironic that</u> he succumbed to a disease that he helped define.

John was one of my first attendings as a pulmonary fellow at San Francisco General Hospital. He was an intimidating presence, especially for a novice fellow. He towered over you, shirt untucked in the back, bowtie in place, and fountain pen in hand, with which he wrote all of his notes. John held all of us to high standards and did not suffer fools lightly. Despite my initial PTSD from rounds with John, I quickly realized that his feedback and guidance helped make me a better doctor. I am proud and honored to be able to call John a mentor and friend over the thirty plus years that I knew him. His death was a great loss for our field, as well as for me.

I was shaken and speechless. Unfortunately, John's death would hardly be the last tragic loss the medical world would experience in the context of Covid.

CHAPTER SIX

The Founding of the FLCCC

The United States declared a national emergency in response to the global pandemic on March 13, 2020. By this point, Paul had posted both an early and a hospital treatment protocol for Covid, largely derived from his HAT protocol for sepsis, on his medical school's website. Each included a number of simple, over the counter treatments and supplements—all of which have since been shown in numerous controlled trials to be effective.

Paul was disturbed to find that the entire country and world had decided that there was nothing to offer in way of treatment for this novel disease. He just couldn't believe it. *No one* was proposing or recommending any type of therapy? It truly was astonishing. (This is hands-down Paul's favorite and most-uttered phrase, even before Covid. Not a day has gone by in Covid that I have not heard him utter, often multiple times a day, "It truly is astonishing!")

Paul wrote a letter proposing a few treatments he felt should be offered to everyone. He sent it to authorities across the country and world including the WHO, Dr. Fauci, the head of the NIH, the head of New York City's Department of Health, and the Health Minister in Lombardy, Italy. He somehow managed to get New York Governor Andrew Cuomo's fax number and faxed him a personalized version of the letter.

March 24, 2020

Governor Andrew M. Cuomo
New York State

Dear Governor Cuomo:

Re: URGENT: COVID treatment protocol. **Patients are dying needlessly.**

I do not think the WH task force are being entirely honest with the American public. Furthermore, the amount of misinformation and the mixed messaging is causing panic and anxiety. While Dr. Fauci is a highly respected scientist, he is not on the front lines and does not take care of critically sick patients.

While it is true that there is no definitive treatment for COVID-19 there is substantial information on treatments that could be of potential benefit. Many of these are FDA approved drugs with an extremely good safety profile. In addition, simple interventions such as Vitamin C, Quercetin, Zinc and melatonin hold great promise in the mitigation of this disease; however as big Pharma will not profit from these agents, it gets no attention. The use of these drugs for COVID-19 are supported by experimental and pre-clinical studies published in the most respected peer reviewed Journals (available on request). In addition, they are cheap and safe; so what does one have too loose.

Dr. Fauci and others are promoting the idea of performing randomized controlled trials (RCTs). I believe that it is unethical to do such trials. How can you offer patients a placebo when testing a drug that you believe may have clinical efficacy? Every patient needs to get the best treatment we can offer; we would expect no less for our loved ones. Furthermore, once these trials are eventually completed we will all be dead, or the pandemic will be over! This does not mean we should not be studying the impact of these interventions; detailed observational studies can provide very useful information.

I am reaching out to you as I pains me to see what is happening in the USA and across the world. While I am usually very modest, this is not a time for modesty. As the most published (and influential) clinician/researcher in critical care in the USA, I believe that I have the scientific and clinical background to understand how to treat these patients and what is possible.
Please see our treatment algorithm attached. I have tried to contact the NIH, CDC, Dr Fauci etc etc. however no one will return my e-mails or calls. I would be happy to discuss this with you personally. I can be reached 24/7 on my cell: ████████.

Kindly,

[signature]

Letter from Professor Paul Marik to Governor Andrew Cuomo, March 24, 2020.

In my hardly humble opinion, he absolutely nailed it.

No one knew how, but within days, Paul's letter to Governor Cuomo was flying around the internet. The issue was, Paul had included his cell phone number in the letter. He began getting calls from all sorts of random people, as well as some high-level folks interested in his protocol. One of them was the commander in charge of Walter Reed National Military

Medical Center; another was a high-level military physician entrusted with formulating a treatment protocol for the Department of Defense.

Paul sent each his protocol and waited anxiously for the enthusiastic, praise-filled replies that never came.

In fact, not a single person or entity who received Paul's letter ever replied. Fortuitously, that letter did get the attention of a woman named Glynis Reinhart, who had grown up with Paul in South Africa. After finding the letter along with Paul's early treatment protocol on his medical school's website, she sent a Facebook message to her friend Joyce Kamen (who would eventually become instrumental to the FLCCC). The message read in part:

> I have sent both you and Fred [Joyce's husband] a letter and protocol for Covid-19 written by Dr. Paul Marik. I grew up with him in South Africa and I think his protocol is worth reading. I am not qualified to critique this protocol but I know that Paul is brilliant and has done incredible work treating sepsis in the past and is very well published. Would be interested to hear what Fred thinks about his protocol.

Joyce is a communications expert with a journalism background and the wife of Dr. Fred Wagshul, a pulmonologist and critical care specialist and medical director of the Lung Center of America. Fred of course was already familiar with Paul's work (you can't be a critical care specialist and not be familiar with Paul's work), and after reviewing the protocol, knew that it made "all the clinical sense in the world."

Joyce called Paul, and he confirmed how well the protocol was working in his ICU patients. Excited and eager to spread the word, Joyce offered to help in any way she could. Paul gave her my name as another HAT expert, and he also gave her the contact for his medical school's press relations office, given the protocol was on the school's website.

A seed was planted.

Right around this time, Paul got a call from Dr. Keith Berkowitz, former partner of Dr. Atkins (of the low-carb revolution) and the founder and director of a prominent internal medicine practice in Manhattan called The Center for Balanced Health. Like many of the founding members of what would become the FLCCC, Keith was an expert in the use of intravenous vitamin C. Paul passed Keith's information on to me, and we talked for hours on that first phone call. Keith was experienced and wise and adamant about one thing above all: Paul and I needed to get our act together and disseminate his protocol. Keith put us in touch with a longtime patient of his, former CBS News correspondent Betsy Ashton, for advice on how to get press coverage.

The very same week, renowned west coast physician Dr. Howard Kornfeld found Paul's protocol on his own medical school's website. Howard is a board-certified emergency medicine specialist best known for his Recovery Without Walls pain clinic in Mill Valley, California. Howard also is an assistant clinical professor at UC San Francisco School of Medicine, founder and president of Pharmacology Policy Institute and an expert in the use of buprenorphine for opiate addiction. I later learned that Howard had organized the conference "Medical Consequences of Nuclear Weapons and Nuclear War," and attended the First Congress of International Physicians for the Prevention of Nuclear War as a full delegate in 1981. As a member of that organization, he co-won the Nobel Peace Prize in 1985.

What I'm saying is that Howard was (and is) a consummate humanitarian and hero.

Howard had been brainstorming with one of his colleagues about how they could help the world with the Covid catastrophe. Their answer: Reach out to Paul Marik and help him disseminate his treatment protocol.

The seed was watered.

In late March, Joyce and Betsy wrote and distributed a press release entitled, "Hospitals use IVs of vitamin C and other low-cost, readily available drugs to cut the death rate for Covid-19 and the need for ventilators." Included were case studies of Paul's treatment of four seriously ill Covid patients, including an eighty-six-year-old man suffering heart disease, who was admitted to the hospital on 100 percent oxygen and not likely to survive. Thanks to Paul's protocol, he lived—as did the other three patients.

Par for what was quickly becoming the course, the media completely ignored the release.

Joyce was not deterred. I personally credit her as the most critical contributor to the FLCCC's early development and survival. Joyce would eventually become our communications director, but it was her early, palpable motivation and commitment which initially (and persistently) lit the fire under me and Paul. Betsy's talents and experience in journalism and TV were also impactful as she would go on to become our creative director and host our popular weekly webinars.

By this point, we—me, Paul, Keith, Howard, Fred, Betsy, and Joyce—were already a team, in the sense that we were in frequent communication with a shared purpose, but we'd yet to formalize our alliance or give ourselves a name. Howard suggested that Paul invite a group of his closest and most prominent colleagues to get more input and support into further developing the protocol.

The seed was beginning to sprout.

Paul recommended me, Umberto Meduri, Joseph Varon, and Jose Iglesias. Umberto, also world renowned, was a professor at the University of Tennessee in Memphis and considered the global expert in the use of corticosteroids in both lung injury syndromes and severe critical illness. Joe was a professor and editor of several medical journals with almost a thousand publications to his credit. Jose was an associate professor, nephrologist, and critical care physician who had recently been the principal investigator of a trial of HAT therapy in severe sepsis. Paul told me that Jose was the smartest fellow he had ever worked with and that he was the only person ever to get a near perfect score on the Internal Medicine in-training exams. Not only was he a walking textbook, but Jose was also considered a master clinician by those who worked with him in ICUs.

On April 4, 2020, Howard wrote to the five of us suggesting that we get together on Zoom to discuss and finalize a consensus approach to treating Covid, with a plan to disseminate this video discussion publicly. That action marked the beginning of what would soon become the FLCCC.

We met via Zoom the next day, and then created an email group where we shared hundreds of emerging papers, preprint manuscripts, and emails with findings and insights from our combined large network of physician colleagues around the country and world. We started having regular meetings to discuss our accumulating knowledge both from studies as well as the care of patients. Together we hashed out a hospital treatment protocol and agreed upon all dosing and combinations of therapies needed. We gave our group a name, the Front Line Covid-19 Critical Care Working Group, or FLCCC (later changed to the Front Line Covid-19 Critical Care Alliance). We created our protocol, which Joyce and Fred worked into an easy-to-recall acronym of MATH+; the MATH part derived from the four key components of Methylprednisolone, Ascorbic acid, Thiamine, and Heparin, with the "+" indicating a few other components including melatonin, zinc, and vitamin D3, which were included based on the high safety, low cost, and emerging scientific data suggesting efficacy.

The seed was a full-fledged plant.

Looking back on the FLCCC's formation and growth, I see it as an improbable series of personal connections and efforts that was founded upon a core belief that we all shared: that this disease could be treated effectively in all its phases with potent, already available, multi-mechanistic therapies. Obviously, there would be no FLCCC without Paul given his singular breadth of knowledge and his ability to build treatment protocols, but Paul would be the first to admit that the FLCCC made impacts far greater than any one of us could make alone. It really did take a village. Joyce in particular was indispensable, encouraging all of us when we were flagging . . . which happened a lot, as it seemed like the entire world (or at least the academic medical system and mainstream media) was against us.

Just because we were a cohesive unit *with a name and everything* doesn't mean we were thriving or even viable, financially. The FLCCC was not an employer to me (yet), but rather it was another commitment in my already unmanageable Covid-ravaged life.

What kept us going is simple: We knew our treatment worked as we could see it working in our patients. We didn't need a randomized controlled trial. We were a group of some of the most widely published and highly experienced critical care doctors in the world, and we had lifesaving information to share. If not us, then who?

Howard donated funds from his nonprofit to hire a website designer named Malik Soomar and FLCCC.net was born. After Howard's funds ran out, I started paying for all the costs associated with the constant editing of our protocol's components on the website. The FLCCC was riding my credit card like a trust fund kid off at college. Then suddenly, an amazing man named Matt Isaacs contacted me to offer a donation of $5,000 which kept the lights on, and the website going, for a while (he later repeated this act of generosity multiple times). But, as the credit card charges continued to mount, and no other funds were coming in, I started getting nervous that I would have to approach my partners, hat in hand, to get them to help pay it off. I was also concerned that if I did that, I might discover that they did not have an appetite to fund the site for the long term.

But I brushed off my fears and decided to just let it ride. From a content and collaboration standpoint, the FLCCC was booming. We were constantly sharing papers and treatment experiences in our group and then discussing them in regular Zoom calls. I knew what we were doing was simply too important to abandon.

We started to take on expert advisors to help expand our breadth of knowledge, including Dr. Eivind Vinjevoll, a specialist in anesthesiology, intensive care, and emergency medicine from Volda, Norway, and Dr. Scott Mitchell, an emergency medicine specialist from Guernsey, Great Britain. Both doctors had expert interest in not only IV vitamin C but all aspects of Covid. Later, during the dreaded Delta wave, I connected with Flavio Cadegiani in Brazil, an endocrinologist who discovered and proved the critical role of anti-androgens in the treatment of severe disease. He was the principal investigator of over a dozen important trials and studies, and he helped add an anti-androgen treatment strategy to our protocol. Next, we met Mary Talley Bowden, an ear, nose, and throat surgeon from Houston who had long been using ivermectin in her early treatment protocols and had successfully treated thousands of Covid patients; for this, the largest hospital system in Texas suspended her privileges and publicly accused her of "spreading dangerous misinformation." It was a gut-punch the rest of us would each experience in turn.

Every day was a torrent of emails between the now-larger group, relentlessly sharing what were hundreds of papers emerging from preprint servers over those first months. Although all of us were working like mad in ICUs, we continued to meet every week or two over Zoom to discuss proposed additions to the protocol as we tried out different strategies and dosing. We were having enormous success with our MATH+ protocol, but despite FLCCC's journalism-savvy and connected members, the media wouldn't touch it. The tiny handful of reporters who deigned to respond explained that we would need a Randomized Controlled Trial (RCT) before they could cover it. An RCT is where participants are randomly assigned to either a treatment group or a control group and is considered by many (but not all) to be the gold standard in studying therapeutics. The reason why I say "not all" is that RCTs in clinical medicine have many problems that are systematically under-recognized or not taught. Chief among their weaknesses is that they are extremely expensive, and the bias of the funders often distorts their conclusions. Plus, they take a lot of time.

We did not have a lot of time. Additionally, another critical component to an RCT is that at the time of the trial, study investigators "must not know which treatment is best."[1] We were already seeing significant success with our protocol, and thus we did not have what is called the "clinical equipoise" to do a placebo-controlled trial. Clinical equipoise is when the investigators leading a study believe that one treatment option is not better than the other; *otherwise, it is unethical to withhold treatment to a placebo group if you know the treatment being tested works.* No one in the media seemed to care about clinical equipoise. They wanted us to withhold lifesaving treatment from some patients to prove what we already knew.

One thing was certain: It was going to be a long, rough road ahead.

CHAPTER SEVEN

My Brush with a Political Legend

One Sunday in April of 2021, I was working an overnight shift in the Electronic Intensive Care Unit (eICU), a place where I manned a huge bank of monitors, remotely managing patients in regional hospital ICUs that didn't have an intensivist overnight. Bleary-eyed between patient calls, I checked my email and found one that had arrived earlier that day from someone named "Thirdwave2." It said (verbatim, with no editing):

> I read that you have found vitamin C useful in treating coronavirus. Is that accurate? newt

I am generally a pleasant, affable, well-mannered person (albeit with a penchant for swearing unfortunately). You can ask anyone who knows me. But it was 3:00 a.m. and I was exhausted and cranky, and I absolutely detest when strangers write to me without a salutation like *Dear Pierre* or *Dear Dr. Kory*. Also, this crackpot hadn't just signed his note "newt;" Thirdwave2 had sent it from "speakergingrich.com." The FLCCC got weird and random emails about our own as well as proposed and speculative treatments all the time, so that wasn't unusual. But this guy was clearly a psychopath. So, I whipped off a rude, semiconscious reply:

> Hey newt. Been up all night in an ICU. Why don't you come at me in a different way. Too many [expletive] wackos. And let's be clear that newt gingrich is one of the biggest morons in the history of the human race. I love this, I am so out of it, I'm about to go to bed, and this schmuck email comes at

me, [expletive] you. Normally I wouldn't respond but you got me at a tender
moment enjoy

I cringe just re-reading that email now (and not just because of my deplorable midnight punctuation and grammar), but that's what I wrote. Again, I received a greetingless reply (which was equally deplorable grammatically, I might add):

After five weeks lock down in Rome I am used to a lot of weird things

I ran across the following quotation

"If you can administer Vitamin C intravenously starting in the Emergency Room and every 6 hours thereafter, while in the hospital, the mortality rate of this disease and the need for mechanical ventilators will likely be greatly reduced, "says Dr. Pierre Kory, the Medical Director of the Trauma and Life Support Center and Chief of the Critical Care Service at the University of Wisconsin in Madison. He explains that it's the inflammation sparked by the Coronavirus, not the virus itself, that kills patients. Inflammation causes a condition called acute respiratory distress syndrome (ARDS), which damages the lungs so that patients, suffering fever, fatigue, and the sense that their inner chest is on fire, eventually cannot breathe without the help of a ventilator."

I thought it would be worth following up on to try to help spread information that would save lives

I did not realize how sensitive you were and how precious your emotions are

Sorry to have bothered you

Newt

So, the lizard guy from Rome was a fan. Wonderful! I deleted his email without replying. The next day, Paul forwarded me this email:

From: Tamera Coleman
Sent: Monday, April 13, 2020
To: EVMS News
Subject: Dr. Paul Marik–Interview Request–Newt Gingrich Podcast
Hello Dr. Marik, Tamera here. I handle the booking for Newt Gingrich's Podcast, Newt's World. It is a non-partisan show where Newt gives historical context behind people and events. We launched last February and have 2.5 million downloads to date with new episodes each Sunday. We are working

on an episode of the podcast centered around "Covid-19 HEALTHCARE" where Dr. Marik would discuss:

- The "HAT" Therapy.

Also, as Newt is located in Rome, the interview will be done via GoToMeeting. Once booked, an email will be forwarded with the GTM link for the connection to Newt. If you do not have computer access, a phone number is included as well.

Looking forward to hearing back soonest, as we would like to confirm booking as soon as possible.

Thanks much, Tamera

Well, damn. Lizard guy was literally *Newt Gingrich*. Keep in mind that I was still a raging liberal back then, indoctrinated by the *New York Times* to hate Republicans in general and Newt Gingrich in particular. He'd put an end to a forty-year Democratic majority in the House. He was the enemy. Still, I never, and I mean *never,* would have spoken to him the way I had in my humiliating email if I'd known who I was writing to. Besides, someone with that kind of media reach could be important in getting information about our protocol not only out to the world but to people high up in government.

I firmly tucked my tail between my legs and set about groveling.

Dear Speaker Gingrich,

Oh boy, this is uncomfortable. It turns out. you are Newt Gingrich? Let me explain my annoyance the other night and trust you can overlook/forgive my actions:

1) I was exhausted, rough night in the ICU
2) I get random people who write to me lately who expect me to write back. but don't offer anything themselves
3) I have a pet peeve when someone who doesn't know me writes to me. but doesn't use a salutation, as in "Dear Dr. Kory or Dear Pierre etc.."
4) The email was from someone named ThirdWave2?? Who is that, clearly some weird email and then the "from newt" was too bizarre/inexplicable and I was too tired to figure it out
5) You were an enemy to my political party, albeit a masterfully effective one which I hate to admit but it certainly doesn't make you a moron

So, I reacted very negatively to this random person who intruded in such a direct and seemingly bizarre way. And it turns out I was writing to Newt Gingrich. Fascinating. Please accept my apology for any offense I caused.

Paul told me you wanted to talk, I would be willing when mutually available. Also, he wants me to put together the rationale for our protocol for the White house. I will tell you this will be the 2nd time my "rationale" for our protocol will get to the White House (Kushner's team). They received it from a very well connected doctor in Manhattan that I am collegial/friendly with (Dr. Keith Berkowitz) and whose very wealthy successful venture capital patient was asked to join the team with Kushner (friends with Kushner?). Berkowitz sent it to his patient who brought it to Kushner and co, Kushners team liked it but then we heard from a commander at Walter Reed who wrote to Paul Marik saying that "the White House very interested in your protocol, but getting push back from the NIH (and later we learned the CDC also pushed back.)

Hope this helps, thanks Pierre

To Mr. Gingrich's enormous credit, he graciously replied. (Do note the salutation!)

Dear Dr. Kory

Thank you for your nice [sic]

I am trying to help get your protocol widely understood and accepted

Two things would really help

1. Your experience at Wisconsin. The University is prestigious enough that a track record there will have some weight.
2. The memo Dr. Berkowitz wrote

By the way I use thirdwave 2 as a tribute to alvin and heidi toffler who were old friends and wrote both future shock and the third wave.

I apologize for being so direct but I write so many different people on so many topics I may get a little too casual

Newt

Wow, he even apologized to me right back. We were becoming buds!

Newt,

Thanks for your explanation, understood. A few things:

1) Paul Marik, cc'ed, asked me to write a one page rationale as per your request. I just sent it to him, he wants to edit it but he is seeing patients so he will need a few hours

2) I wrote the rationale that Berkowitz gave to his patient on Kushner's team. It is longer, more detailed, but more "evidence-based" i.e. with publication references etc. I think the one Paul is editing is better/more focused. I can send both

3) My experience at Wisconsin will be hard to relay accurately or quickly. These data are hard to gather as I personally only treated 2-3 patients and treatment was begun very late in the disease course as they were transferred from the region. I have largely been performing leadership, not clinical functions over past month.

4) However, I did succeed at getting it on our Univ. of Wisconsin treatment guidelines and thus the "hospitalists" on the wards were using it, prior to their arriving in the ICU, which is critical to the effectiveness of our protocol at avoiding ICU admission and need for a mechanical ventilator. I can try to get these numbers in order to say "x were treated early, and y required ICU admission" with the hopes it is an impactful number. This will take time, I need at least a day, chart reviews are laborious. Thanks, Pierre

Newt replied:

Pierre

Thank you very very much

It will be a great help if you can pull that together

I am following up with paul and will be pushing this out including going back to the white house newt

And pull it all together I did. I sent Newt my entire bag of tricks, with papers, emails, links, and lengthy explanations to support it all. I hit send and sat back with a satisfied, hopeful smile. We had a political heavy-hitter on our side! (A Republican one, but still.) As "The Jerk" Navin Johnson would say, *things were going to start happening to us now.*

You likely know by this point in this book that of course, things did not start happening to us then—at least not in any public or profound way. Paul appeared on Mr. Gingrich's podcast and it went quite well, after which my buddy Newt and I continued to correspond. He wanted a summary of our protocols to share with his colleagues and I gave it to him. I can only imagine he was laughed out of the chamber (or cigar bar or chat room), as I don't think I heard from him again after that point. By then, Fauci was firmly installed at the helm, and the Official Messaging at the time essentially amounted to, "isolate until you're on death's door and then get yourself to a hospital and hope for the best."

It was the Cuomo Show all over again, but we weren't daunted. We knew going in that this was going to be an uphill slog; now it was clear we were going to be huffing up that hill in a hurricane, carrying angry porcupines and wearing banana peel slippers.

CHAPTER EIGHT

Fighting for Air

Nearly every Covid rabbit hole I have gone down eventually led to a public science battle, only some of which I have won. Of the few I emerged victorious (like corticosteroids), none was more important than when I was able to shut down the shocking, proliferating practice of ER and ICU doctors putting Covid patients on ventilators unnecessarily and dangerously early.

As a clinical leader in pulmonary and ICU medicine at UW, I was one of the more experienced ICU clinicians. I was also known as a "vent geek." In fact, one of the reasons I became a pulmonary and critical care doctor stemmed from an early fascination with operating mechanical ventilators. Subsequently, I have long taught the management of acute respiratory failure and mechanical ventilation to medical students, residents, and fellows. One of my core teaching points has focused on identifying the optimal timing for the decision to transition a patient to a mechanical ventilator.

Guidance on how to make the decision is simple conceptually but stressfully complex in practice. Basically, the optimal timing of transition to mechanical ventilation is not too early . . . and not too late. See how simple that is?

The reason for this approach is that mechanical ventilators are double edged swords in that they can absolutely be lifesaving when truly indicated (and thus, the benefits outweigh risks), but they also can injure the lungs when used inexpertly or prematurely.

The deleterious effects of mechanical ventilation arise from the fact that it often requires prolonged sedation and immobility which can cause confusion, delirium, muscle atrophy, and weakness. All of which delay recovery and open patients up to developing complications. Invariably, the less time you spend in an ICU, the better you will do.

In other words, the timing of the decision is critical. Do it too early and you likely will be doing it unnecessarily in a proportion of cases; doing it too late leads to a procedure with higher risks, as the act of intubating someone in severe distress with low oxygen is much riskier than in a more stable patient. Knowing precisely when to intervene when a patient's respiratory status is deteriorating is a critical and challenging patient care issue. In simple terms, you are sedating and paralyzing someone in order to insert a breathing tube through the vocal cords and into the trachea, a procedure that presents a rare but catastrophic risk.

If you don't establish a supportive airway quickly in some patients, a cardiac arrest can ensue. Fortunately, due to modern intubating techniques, equipment such as video laryngoscopes, simulation training practices, and sedation and paralysis protocols, death is rare . . . but still not an impossibility. And although death is quite uncommon, I have been involved in more stressful intubation scenarios than I (or my patients) would have liked. "Managing a difficult airway" is the emergency of all emergencies, as you are trying to prevent a cardiac arrest from deprivation of oxygen and/or excessive respiratory fatigue.

If I was going to err on one side or the other, it would always be *too late*. I would try to give every patient as much time and treatment as I could until it was clear they were not improving dramatically or quickly enough to avoid mechanical ventilation. I tried to give them every possible chance without endangering them. So, I would consider myself a "late intubator" by practice. The comfort level with deciding on the appropriate time to intubate obviously varies across physicians as their risk tolerance—and their perceptions of the competing risks—varies according to their training, experience, and personalities.

As Covid patients began to be admitted to UW Hospital, a number of my colleagues approached me and suggested that we institute a rule for when to put someone on a ventilator. They proposed using the amount of oxygen a patient was requiring as an indicator. I thought this was insane but I also understood where it was coming from: the doctors were scared, as they had not developed familiarity with the disease. This fear of course was compounded by rumors or reports of Covid patients who were supposedly coming in with low oxygen levels and who, despite oxygen supplementation and looking fairly stable, would suddenly "crash" and require emergency intubation or resuscitation.

Although I believe these doctors were advocating for early intubation largely for the safety of the patient, I knew this would paradoxically spell disaster if the practice became standard. Plus, I had serious doubts that pneumonia or pneumonitis would cause such sudden crashes.

I have spent my career consulting on patients in various forms and degrees of respiratory distress, and all respiratory failure conditions tend to have a general trajectory and response to medication. Because of this, knowing when to intubate becomes easier as you gain more experience. In my first job after fellowship training, my hospital was poorly staffed with pulmonologists and intensivists, and I saw double the number of patients compared to the average full time ICU doctor. I worked sixty to eighty hours a week, plus I frequently moonlighted overnight. I gained a ton of experience (and expertise) fairly quickly.

In those early Covid days, I refused to believe that an inflamed lung would lead to a precipitous crash. I knew this intuitively, but also from talking to my colleagues on the front lines in New York City. I argued with the early intubation crowd that, even though this was a novel disease, it didn't change the foundational principle of when to institute mechanical ventilation.

At one of the daily briefings that I led at UW, which was attended in person and remotely by all residents, hospitalists, and intensivists in charge of taking care of Covid patients, I argued that the indication for the institution of mechanical ventilation should never be based on an oxygen level and instead must be almost solely based on an assessment of the patients "work of breathing" [basically the energy necessary to inhale and exhale] and their subsequent ability to sustain it. This is where it gets a bit more complicated, as a patient's ability to sustain an elevated work of breathing is itself dependent on multiple factors such as their frailty, mental status, and the cause of their respiratory failure (some conditions are more easily and quickly reversed than others).

My colleagues and trainees carefully listened and for maybe the last time in the pandemic, trusted my judgment without much argument. *The idea of setting arbitrary oxygen limits as the trigger for intubation simply disappeared.* I'm pretty proud of that because I know it was not the case around the country. Many hospitals and academic medical centers were using arbitrary limits to put patients on ventilators, and I believe this was one important factor which led to the widespread need for additional ICU rooms as well as ventilator shortages.

As inappropriate and unnecessary as early intubation was, I believe that lack of treatment caused far more unnecessary deaths than ventilators did. The "when to intubate" conversation would have been unnecessary if patients had been appropriately and aggressively treated with corticosteroids and blood thinners. When people say, "a ventilator killed my grandmother," I want to say, "ventilators don't kill people; cowardly doctors do."

Fortunately, the early intubation practice was phased out fairly quickly in most hospitals as doctors gained more experience in managing Covid patients. They began to recognize that the pulmonary phase of Covid presented as a relatively unique form of respiratory failure, in that patients would come in with often quite low blood oxygen levels yet would appear fairly comfortable in terms of their work of breathing, a condition doctors began calling "happy hypoxia."

Many doctors began using high flow oxygen devices instead of mechanical ventilation. These devices, called *heated high flow nasal cannulas* (HHFNC) are a marvel of technology as you can deliver incredibly high quantities of oxygen (up to sixty liters per minute) into a patient's nose as the oxygen is 100 percent humidified and heated. With normal low flow nasal cannulas, which are not fully humidified or heated, if you try to increase the flow past five liters per minute, patients cannot tolerate it due to discomfort and dryness. HHFNC became the workhorse of Covid, and I believe many lives were saved by those devices.

I can't speak for the rest of the country or world, and certainly some places would have adopted more conservative approaches before others, but by August 2020, the hospital where I was working was using heated high flow nasal cannulas aggressively in lieu of ventilators. Any doctor worth the paper his or her degree was printed on knew that you could support a Covid patient on an HHFNC for days to weeks before needing intubation.

And therein lies the problem. If the average person knew how bad the average doctor is, they'd probably never go to a hospital again. The reality is most doctors are terrible at using judgment and critical thinking. They like rules and protocols and guidelines and can cite them all on command, even when different approaches make more sense. In my entire career, I would estimate that maybe ten percent of doctors I've worked with are the critical-thinking kind, and that's being generous.

It also had everything to do with the spiraling Covid crisis we had on our hands.

CHAPTER NINE

I Lose My First of Three Jobs in the Pandemic

After weeks of preparing ICU rooms and creating complex ICU specialist team schedules in preparation for a surge of critical Covid cases, patients began to trickle into UW. The trickle soon turned to a steady stream. Given that Covid was so new, practitioners were scrambling to learn everything we could about the novel virus—and more importantly, how to counteract it. The latter attempt is where my first troubles began.

Working closely with colleagues in the FLCCC as well as in ICUs in New York, Italy, and China, we learned that both corticosteroids and anti-coagulation were critical to survival. The optimal dose, drug, timing, and duration for these two strategies was not yet known, but the need for them was undeniable.

The problem was that the combination had not yet been "proven" in some large, prospective, multi-center, double-blind, randomized controlled trial, and that is currently the only evidence that can make changes to therapeutics in the US health system. The horrific departure of this policy and practice from the long-standing reliance of physicians on the powers of medical knowledge, logic, observation, reason, pragmatism, and the precautionary principle of relying on risk/benefit assessments is now legion. It also quickly led to the most catastrophic mortality rates of ICU patients in history. All because the entirety of US academic medicine, over the last twenty years, has been reduced to a "Church of RCT Fundamentalism," a conversion that was wholly fueled by Big Pharma, as they essentially control the funding, design, and even outcomes of such trials.

This approach was catastrophic. Not only were centers running out of ventilators to support patients, but even with enough vents, patients were dying in droves. One hospital system in New York reported an 88 percent mortality rate among patients on mechanical ventilation. This was unprecedented. And these patients were dying because with few exceptions, they *were not being treated* with anything beyond oxygen, fluids, and Tylenol. The "official" policy emanating from nearly every hospital therapeutic committee in the country was to largely not recommend anything outside of a clinical trial.

While I was getting serious pushback for recommending the use of anticoagulation and corticosteroids, UW, like a majority of academic medical centers, was embracing the use of convalescent plasma, which is blood from people who have recovered from an infection. Yes, they were enthusiastically soliciting blood donations from the Covid recovered and transfusing that plasma into the newly ill. It was insane. There was no evidence for the efficacy of this therapy outside of bloodstream infections, and Covid was a respiratory virus. Not-so-fun fact: Since that time, all of the studies on convalescent plasma have been compiled and show that it leads to an increase in mortality.

Prior to Covid, a doctor could, for example, treat patients with medicines that had been approved for a different illness but that hadn't been approved specifically for the illness you are treating, a practice known as off-label use, or the "repurposing" of a drug. Different diseases can cause illness using similar mechanisms, thus a medicine that works in one disease can also work in others. Further, some medicines have multiple mechanisms of action, thus they can be effective in more than one disease. Doctors could rely on their own judgment and experience and knowledge of a medicine's actions to make treatment calls that may not (yet) be backed by peer-reviewed studies. They could think critically. They could be doctors.

And now, the powers that be were prioritizing medical research ahead of the welfare of patients. They wanted control groups—which would mean simply not treating some patients at all. We doctors wanted to try anything, everything, that had even a sliver of a chance of saving lives. If it was your parent, child, or spouse dying in that hospital bed, would you be on the side of medical research . . . or medical ethics?

A *New York Times Magazine* article quoted one researcher as saying that relying on gut instinct rather than evidence (in other words, published RCTs) was essentially witchcraft. Witchcraft! The article went on to quote a bioethicist at New York University's medical school, referring to the early and widespread off-label use of hydroxychloroquine in Covid patients. "Panicked rhetoric about right-to-try must be aggressively discouraged in order for scientists to learn what regimens or vaccines actually work," the so-called expert insisted.[1] In other words, they needed a control group. Keep in mind, this is a person whose position is centered around *medical ethics*. It was astonishing.

I was heavily profiled in that article, the highlight of which was when the editor of the *New England Journal of Medicine* called me "lucky" for predicting the critical need for corticosteroids in Covid patients. My colleagues in the FLCCC would call me that multiple times over the next few years as we kept getting everything right.

Yup. *Lucky Pierre*.

The article took the classic "presenting both sides" approach, leaving readers to decide which therapeutic approach you would have wanted as a patient in early Covid. Keep in mind, this was way before Paul Marik and the FLCCC had identified ivermectin as a highly effective treatment.

I left UW fairly quickly after the first wave of patients arrived in late March, propelled by two particularly disturbing incidents.

Disturbing incident #1: As the Clinical Service Chief of the Critical Care Service at the time, I led daily clinical webinar conferences with as many as fifty hospitalists, ICU specialists, and residents. These were essential given the deluge of new information, changes to policies, and other developments unfolding daily. Surely due to my subtle and not-so-subtle advocacy for empiric use of medicines and my avoidance of early intubation, which many were endorsing by this time, my superiors gradually took over the leading of these calls. As you can imagine, their guidance was wildly different from mine and centered on recommending "supportive care only." You know, because it was a viral syndrome, and we didn't have any effective antivirals for coronaviruses (or so we allopathic doctors had been trained to believe). These were not patients sitting home with sniffles and a cough, mind you; they were suffering severe organ failures and dying. I was trying to save lives; the folks running the show were saying, "let's keep you comfortable while you die."

It quickly became clear to me that my clinical opinions were no longer desired.

The sad part is, during this time I was in constant contact with colleagues from around the world. We shared ideas and success stories in messaging apps and on social media (before such discussions were censored and banned there). More than a few colleagues were using corticosteroids in Covid patients with remarkable results. One memorable social media post from a physician in New Orleans read, "We floundered for two weeks. Lots of codes, intubations, and death. Maybe 15 discharges. We started steroids and discharged 250 patients. Less intubations, less codes. And the ones that ended up on vent, not as serious." Another wrote, "Since we started using steroids, we were able to free ventilators and get elderly patients out of the hospital without needing a ventilator. Patients who were crashing quickly, who we had to have end of life talks with, were able to walk out of the hospital. At no point did any of our patients worsen because of steroids."

These were invaluable reports from the front lines—or so we thought. But since this info emanated from social media and was not approved by the alphabet agencies or beloved journals, it fell into the category of what would later be attacked as *medical misinformation*. The authorities continued on with their moronic "supportive care only" approaches. ICU bed numbers continued to swell. People died. A lot of them. Often after weeks of being in a hospital or ICU, alone, not able to spend their final moments with their loved ones. It was both unforgettable and unforgivable.

Disturbing incident #2: One of my superiors went behind my back and convinced my fellow members of the Covid therapeutics committee to remove intravenous vitamin C from the University of Wisconsin (UW) guideline, after I had successfully and scientifically advocated for its inclusion based on numerous studies. This was academic misconduct which would threaten the lives of patients coming to UW. It also was done on behalf of the dean of the medical school at the time, Dr. Robert Golden, a psychiatrist who had become incensed when he read a newspaper interview where I mentioned that high dose intravenous vitamin C would likely be critical to improving survival in Covid. As soon as I discovered this insanely corrupt action, I requested a leave. Note that at the time, the use of intravenous vitamin C in ICU patients with severe respiratory failure was actually supported by a large RCT, published in a major medical journal,[2] which found it led to a massive reduction in mortality in ICU patients with severe lung injury. More recently, a meta-analysis of IV vitamin C trials in Covid[3] amassed evidence strongly suggesting numerous benefits, including survival, especially in the most severe patients. "Lucky Pierre" strikes again. Moral of the story: This is what happens when psychiatrists decide to dictate the care of ICU patients while servile, ignorant physician leaders defer to such insanity.

As a native New Yorker, I badly wanted to fight on what was at the time the medical front line of the United States battle against Covid. My email inbox was filled every day with bold headlined emails from all the critical care societies: *Critical Care Specialists Needed in New York Urgently*. I *had* to go. However, in my mind, I had already decided that I was more than just asking for a leave; I was formulating a plan to resign once I got to New York. My wife supported me from the beginning, reminding me how I had become increasingly miserable in the Ivory Tower during my five-year tenure as Critical Care Service Chief.

This is the email I wrote to my boss asking for the leave:

Humanitarian Leave – I would like to request permission for a leave, for several reasons:

A. My approach in advocating for a strategy of care for critically ill Covid patients has created discord and tension at a time when consensus and unity are critical. I cannot see how I can be a force in achieving those aims when my clinical judgment and recommendations are in direct opposition to that of divisional, departmental, and Medical School leadership.

B. I feel my skill set and effort would be most valuable and have maximum impact if I were to be allowed to support hospitals in New York City that are overwhelmed with critically ill Covid patients.

C. Please note that I am not formally scheduled on a critical care team until the first week of May and I would be willing to return to fulfill that responsibility if desired. Further, I feel that UW Hospital is currently well prepared to care for any surge of patients at this point and my hopes are that an overwhelming surge is being avoided by early and current social distancing practices.

D. Medical Director/Service Chief positions – I would like to offer my resignation from these roles. It has become clear that I am not suitable for these roles. They are just not in my skillset or personality. I deeply apologize for my actions in the past week, actions which were solely intended to rapidly develop an effective strategy of care for Covid patients. The moral distress that I have been suffering over what I "too strongly" [his words to me] perceive as ineffective and inadequate care and needless death has overwhelmed me.

The next night, after a brief and strained final remote conference meeting with my soon-to-be ex-Chair of Medicine, she granted me permission for the leave (I suspect gladly). I bought a plane ticket to LaGuardia, and off I went.

CHAPTER TEN

My Senate Testimony on Steroids

I got to Mount Sinai Beth Israel Medical Center in lower Manhattan during the latter half of that insane first surge of patients that overwhelmed all of New York City's health systems. Within a week of taking over my old ICU, I received a call on my cell phone from the chief of staff of Senator Ron Johnson, then Chairman of the Homeland Security Committee. Apparently, he wanted to talk to me about our FLCCC protocol for treating hospitalized patients. Senator Johnson was uniquely and rightly concerned about how doctors were failing at identifying and employing both early phase and late phase treatments while so many patients were dying.

Senator Johnson told me that he had decided to hold hearings as to why, in my words (and likely his, too) "the doctors were not doctoring" and US citizens were effectively being denied treatments or attempts at treatment. I was invited because he had found the FLCCC website and protocols and noticed that one of the founders was from Wisconsin. Watching that hearing again a year and half later, I was struck by how much we knew back then about how to treat the disease and how strongly we made the argument for corticosteroids.[1] Months later, all the randomized trials would validate their use.

Senator Johnson was pleasantly shocked to find a mature treatment protocol using a combination of effective therapies (with each component later shown to be effective in clinical studies). From my memory of our first conversation, he told me that the underlying reason why he was trying to bring attention to the issue was because he wanted doctors to "take their

$%&#! gloves off." (I actually don't think he swore, but I recall that he said it extremely powerfully.)

I delivered that testimony remotely from my tiny "emergency volunteer" hotel room paid for by Mount Sinai, and at the time, my words were quite provocative. Not to mention effective. Ron's hearing surely saved a lot of lives as he heard from doctors for months afterward about how my testimony emboldened them to try corticosteroids and as a result, were able to report numerous robust recoveries. I'm extremely proud of that. My admiration for Ron grew as I got to know him; he is truly an intensely caring man and committed senator. In what would have shocked me to my core three years ago, a career highlight for both me and Paul was being invited to his election night party and spending time with him and his wonderful family.

Although my sworn statement in that hearing did not reach as many ears as my later and more widely disseminated ivermectin testimony,[2] my steroid testimony got the attention of the University of Wisconsin—and not in a good way. Although I had already resigned, my resignation was dated for June 30, which was two months away. I still held my academic title of Associate Professor of Medicine (in fact, I was on the brink of becoming a full professor when I resigned), so I was still being quoted as an associate professor at UW in the press.

My boss was livid, his boss was livid (she was the Chair of Medicine whose academic misconduct led to my resignation), and *her* boss (the vitamin C–hating psychiatrist dean of the medical school) was livid.

Why were they so furious? Because an associate professor of the Medical School had spoken publicly. Without telling them first. Without asking for permission. Without alerting their press office. And with scientific opinions that ran counter to not only those of the Covid Therapeutics Committee of the University of Wisconsin but also the NIH, FDA, and all other national and international health care societies at the time. The First Amendment's a bitch.

Note that I wrote above *at the time*. Eight weeks after that hearing, the use of corticosteroids became the standard of care worldwide for hospitalized Covid patients after a large study in the UK showed that *the use of corticosteroids led to dramatic reductions in mortality.*

But at the time, I was an outlier; a charlatan. This is what the "medical community" thought about the use of corticosteroids in the immediate aftermath of my testimony:

Public health agency and professional society guidelines on corticosteroids

Source	Recommendations
Australia June 3	Avoid corticosteroids in people with suspected or confirmed COVID-19 unless there is an evidence-based indication for them (e.g. severe acute exacerbation of COPD or asthma). (based on expert opinion) CEP NOTE: this guidance is for patients with moderate to severe disease; guidance for patients with severe to critical disease do not include recommendations for or against corticosteroids.
WHO May 27	We recommend against the routine use of systemic corticosteroids for treatment of viral pneumonia.
CCCS May 19	In patients with severe COVID-19 and ARDS we suggest using corticosteroids (Weak recommendation). In patients with severe COVID-19 who do not have ARDS we suggest NOT using corticosteroids (Weak recommendation).
DoD May 14	There is a strong consideration to avoid routine steroids based on early data out of China as well as other studies related to Middle Eastern Respiratory Syndrome Coronavirus (MERS-CoV) which have shown that steroids delay viral clearance. However, new consensus guidelines recommend considering methylprednisolone for intubated COVID-19 patients with ARDS.
NIH May 12	For critically ill patients with COVID-19: The Panel recommends against the routine use of systemic corticosteroids for the treatment of mechanically ventilated patients with COVID-19 without acute respiratory distress syndrome (ARDS) (strong recommendation based on expert opinion).

Figure 2. Health agency and professional society recommendations on the use of corticosteroids in Covid-19, June 8, 2020. (Source: Center for Evidence-Based Practice, http://www.uphs.upenn.edu/cep)

But I continued to do what I was doing. At the end of each indescribably and horrifically difficult day in Beth Israel's main Covid ICU, I was getting a call from my superiors. One repeatedly insisted that I needed to alert the University's media relations office before I spoke to any more press while also telling me that my opinions were likely dangerous. A UW media relations employee had discreetly told me after I was publicly quoted in an article recommending intravenous vitamin C, that "although the Chancellor of the University fully supports a faculty's ability to speak their opinions freely, *the Dean of the University's School of Medicine* has a different opinion." I became increasingly annoyed at these calls, and finally got so pissed off that I hired the best civil rights and employment lawyer in the state of Wisconsin, Attorney Lester Pines.

His letter to them was an absolute tour-de-force. I loved it so much, I actually had it framed. It hangs in my office today. Please enjoy:

May 19, 2020
Re: Dr. Pierre D. Kory
Dear Dean Golden and Dr. Schnapp:

I represent Dr. Pierre Kory. I have reviewed the multiple email correspondences that he has received regarding public statements and public appearances about the treatment of Covid-19 patients that Dr. Kory has made and will make.

Dr. Kory, while on his own personal time, is free to make any public statements or express any opinions he desires about the proper way to treat patients who are critically ill with Covid-19, or on any other medical matter. When Dr. Kory makes such statements or provides such opinions, he is not speaking on behalf of the University of Wisconsin School of Medicine & Public Health ("SMPH"). Nor is he speaking on behalf of UW Health. He has not at any time suggested that he is.

The May 18, 2020 letter told Dr. Kory that before making public statements or public appearances he is "required to inform and work directly with Connie Schulze, Director of Governmental Communications for SMPH and UW Health" and that he is expected to also "inform Dr. Lynn Schnapp, department chair and Allison Golden, the Office of the Dean." There is absolutely no requirement that a faculty or academic staff member at the University of Wisconsin, including the SMPH, or on the staff of UW Health must work with the Director of Governmental Communications or anyone else before testifying before Congress, publicly appearing before any other public organization or agency, or making a public statement. Nor is there any requirement to inform either a department chair or the Dean.

Not only is Dr. Kory free on his own time to publicly share his knowledge and opinions, he is obliged to do so. As a faculty member of the SMPH and the University of Wisconsin, Dr. Kory is expected to ensure that his medical knowledge is shared beyond the boundaries of the campus. Neither the SMPH nor UW Health has the right to prohibit Dr. Kory from doing so, regulate what he may say, or where he may say it.

As a public employee, Dr. Kory's rights are protected by the First Amendment to the United States Constitution and Article I, Section 1 of the Wisconsin Constitution. When he is speaking or writing on matters of public concern and doing so on his own time, he may speak or write without interference from or control by his employer. He is also free to identify himself by his medical and academic credentials and will continue to do so.

Calling and emailing Dr. Kory repeatedly to express displeasure for his public expression of expert opinions, baselessly accusing him of trying to represent his opinions as those of the SMPH, and then claiming he is required to inform and work directly with University representatives prior to speaking publicly, are attempts to intimidate and harass Dr. Kory and restrict his free speech rights.

Multiple actions taken by Drs. Schnapp and Dean Golden have been unfounded, political attacks on Dr. Kory's credibility and expertise. I suggest that all of you refrain from any further such behavior. I strongly urge both the SMPH and UW Health to refrain from interfering with Dr. Kory's right to provide information to the public during this time of urgent need for the best clinical care possible for Covid-19 patients.

Dr. Kory is a highly respected clinician who has been hard at work on the high-risk front lines of the Covid pandemic. It is unconscionable that he be forced to endure repeated intrusions during his work and advocacy. While there may be those who disagree with his approach to treatment, they should pay attention to the scientific evidence on which it is based rather than try to suppress his public discussions about it.

Very truly yours,
Lester A. Pines
PINES BACH LLP

It wasn't just a masterpiece; it also got the harassing phone calls and emails to stop. I treasure my framed copy of it, as it's a powerful reminder of what academic freedom is supposed to look like.

CHAPTER ELEVEN

The Spring Surge

When I had first arrived at Mount Sinai Beth Israel to take over my old ICU, I received a warm and grateful welcome by Charles "Cap" Powell, the chief of Critical Care at Mount Sinai. (When I'd left Beth Israel in 2015, Cap had wanted to recruit me to head the ICU uptown at the main campus hospital of Mount Sinai). He hosted a "meet and greet" with hors d'oeuvres and drinks for me and the other emergency volunteers who had arrived at the same time. (I should add that we weren't technically volunteers since we did receive a modest salary but still, we were there voluntarily.) Although the hospitals and ICUs were still overwhelmingly busy, hearing the war stories from the early weeks of the surge made me realize that I had missed the worst of it. Things on the ground were now somewhat less chaotic and better organized. The hospital had acquired significantly more resources to manage the patients, both in terms of physician and nurse staffing as well as equipment such as oxygen and cardiac monitors for the still overwhelming numbers of unstable patients.

I checked into the hotel near the hospital where the other emergency volunteers were staying. The room was tiny even for Manhattan standards, with maybe eighteen inches of space between the king bed and the walls. The sink in the bathroom was so small that I could barely rub my hands together under the faucet. But there was a miniature desk which I used during the majority of my free hours to support all the papers and op-eds I was writing.

After getting settled, I went to visit my parents who lived in an apartment building ten blocks up and one avenue over. This became my routine:

work all day, go to the hotel and change, and then walk over to my parents for a home-cooked meal. (My mother is French and an amazing cook.) They had set up a small, portable table across the main room where I would eat alone, while they ate at the dining table ten feet from me. I know. It was stupid—something I even felt at the time, but my parents were more comfortable at a distance from me. I had missed seeing them regularly since I had moved to Wisconsin, so we all enjoyed these nightly dinners together, something that hadn't happened since I was a kid.

The next day, I showed up for work at Beth Israel, again with a relieved and warm welcome from my former colleagues, most of whom I had either trained as fellows or had collaborated with over the years in teaching ultra-sound both regionally and nationally. They were battle-scarred "Covid veterans" at that point and were and are some of my favorite doctors: high-level clinicians and clinical researchers and old friends like Jason Filopei, Young Lee, Paru Patrawalla, Lena Minakawa and others. They were led by the Chief of Division David Steiger, who started his position the day after I left in 2015. I developed a fondness and respect for him during his recruitment, and would have loved to have worked for him, but 2015 was when I was burned out—and David understood. In fact, after I left, he took on my father as one of his outpatients which was hugely reassuring to me from Wisconsin.

David and I began exchanging insights and approaches and had a great discussion on the use of corticosteroids. I was not surprised to discover there had been some pretty fierce fights between the infectious disease docs and my ICU colleagues on this issue during the surge. It was the same story as with my old fellow in Detroit and the horror show at UW. Of course, I arrived with steadfast support for the use of corticosteroids which my colleagues had already started to use over the objections of the ID doctors (however not in a standardized fashion). It was a relief to find that my guidance and insights were openly welcomed, something I had not experienced at UW. With each reasonable discussion, the more I felt that when I left UW, I had escaped from a cult.

After sharing all of my supporting data, corticosteroids became somewhat of a standard treatment there—at least, it seemed to me—even seeing increased use by the hospitalists, who were traditionally much more apprehensive than us intensivists.

They assigned me to run my old Medical ICU which was now the main Covid ICU. It was full and everyone was on a vent. Even worse, almost the entire unit had been on a ventilator for weeks. It was the first ICU to fill with Covid patients in the early days of the surge, and those who hadn't died were still alive . . . but barely. Almost all were on maximal ventilator settings, but that wasn't their only problem. Several had kidney failure and most had severe encephalopathy (showing little or severely altered brain function). It was a heartbreaking sight.

I led a large team of residents and taught them my insights and approaches to Covid. I put every patient on MATH+ at high doses. A few showed modest improvements, but for the majority it was started too late in their course to have an impact. I coordinated care with both the Covid intubation team staffed by out of work anesthesiologists and the tracheostomy team staffed by out of work ear, nose, and throat surgeons (at that point, all non-emergent surgeries had been canceled along with outpatient practices). Patients who were on non-maximal ventilator settings or showed some lung function improvements received a tracheostomy, a more stable, comfortable way to be on a ventilator that involves a short tube inserted directly into the trachea through the neck instead of the tube passing through the mouth and vocal cords into the trachea.

As had been the case throughout most of my career, my days were split into what amounted to being an intensivist in the morning and a palliative care doctor in the afternoon. Mornings were spent evaluating and responding to varying and multiple degrees of organ function and afternoons were spent in family meetings. Since no relatives were allowed into the hospital, I had each resident on my team give daily updates to patient families by phone, while for the unstable or deteriorating patients, I held longer family meetings using an iPad with video conferencing software so all family members could attend. We had a team solely responsible for organizing and scheduling these meetings. Family support has always been a core objective in the care of my ICU patients, but here it was harder because the families could not be physically present. So we did what we could to stay in constant contact and support.

In most patients, I was doing CT scans (much more sophisticated than chest X-rays, although still not without limitations) to help with prognostication. Almost all were in advanced degrees of severe pulmonary fibrosis or scarring and were showing continued deterioration over time. Since my colleagues had seen a couple of patients improve and survive from such an advanced lung failure, we were cautiously optimistic when we could be.

Those four weeks were the hardest in my ICU life.

One of my patients was a fifty-year-old man (same age as me at the time) I'll call Jose. Jose, also like me, was the father of three girls. He was on maximal support via a "non-invasive" ventilator, meaning pressured air was delivered by a tight mask strapped to his face instead of via a tube in his trachea. He'd been on that machine around the clock for eighteen days when I took over his care. I don't think most non-medical people can understand how abnormal that is, given that non-invasive ventilators are traditionally used either intermittently or short-term. Enduring eighteen days of continuous pressurized oxygen via a tight face mask was a remarkable feat of stoicism and strength I had never seen in my career.

I treated Jose with everything I could but there was no change in his lung function. He was rapidly losing weight because he couldn't take more than a few sips of a protein shake at a time; every time he removed his mask, his oxygen would plummet, and he would start to turn blue. But he didn't complain and simply sat in his bed, lying prone on his stomach as much as he could to improve oxygenation. He was literally fighting for his life.

On day twenty-six, I decided to recommend Jose for a lung transplant, something which is almost never done in acute respiratory failure as it is intended for those with advanced chronic diseases only. However, since he wasn't on an invasive ventilator, I hoped they would consider.

There were essentially three lung transplant centers in the city. Two told me that they weren't considering such patients at the time. However, when I called Montefiore, the lung transplant head there was much more open. We had a long discussion and he agreed to consider Jose for transplant. When I told Jose (and his family) the plan, his whole demeanor changed. He even smiled, and I saw him fighting even harder to breathe on his ventilator.

The problem was, to get Jose a formal, comprehensive evaluation for transplant, we had to get him up to the North Bronx from lower Manhattan. This was no easy task, as transporting someone across the city on maximal support incurs significant risk. We made sure to get an advanced paramedic team with experience in such high-risk transports. I was excited because this was likely going to be the first lung transplant in an acutely ill Covid patient, not only in the country but in the world (the first reports of successful transplants, although few, would be reported in the coming months). I will never forget the spirit of my ICU fellows, residents, and nurses, but especially Jose and his family. He gave us all a thumbs up sign as the paramedics wheeled him out of the ICU.

You have no idea how badly I want to tell you all about Jose's miraculous, groundbreaking recovery. But that's not what happened.

As I would soon learn from Monte's transplant chief, tragically, Jose—a husband, father, fighter—died within days of arrival due to a complication which occurred during that transport. The noninvasive ventilator that they used for the transport malfunctioned and he arrived in severe distress, barely conscious, with extremely low oxygen levels. They had to put him on an invasive ventilator during which he suffered a pneumothorax (punctured lung). His blood pressure plummeted, which led to kidney failure, shock and ultimately, death.

I was devastated. I broke the news to the entire team during our morning rounds and we held a long, deeply somber moment of silence. My only consolation is that I believe Jose and his family knew that I had done everything in my power to give him a chance to survive.

At the end of four weeks, my ICU was empty of Covid patients. The wave had crashed, and case numbers were down to the point of allowing us to cohort all the new and old COVID patients in the other ICUs that had been created. My ICU had gone back to being the general medical ICU. In the other units, although there were still new admissions, all the existing Covid patients had either died or had received a tracheostomy and were transferred to "step-down" units or facilities for prolonged weaning off ventilators.

Things were much more stable on the ground by that point, so I returned to Wisconsin. My daughter was still quite sick, and I was practically a walking petri dish, so I stayed in a hotel locally for a few days. I needed to be sure I wasn't going to bring the dreaded, deadly disease into my home.

CHAPTER TWELVE

Happy Hypoxia

When I was surrounded by Covid patients in New York, I was intrigued by the number of patients presenting early on with "happy hypoxia," a description for the unusual combination of severely low oxygen levels yet often without extreme breathlessness or discomfort. I had seen this kind of presentation on occasion before, but it was rare, and I couldn't put my finger on exactly which disease or situation I had seen it in. One night, while obsessing about it as I was trying to fall asleep, it hit me that Covid patients reminded me of patients I'd seen with a rare-ish disease called *organizing pneumonia* (OP). The name is actually a confusing one because OP is not an infectious illness, but instead a reaction to a lung injury, typically from an unknown source or most commonly a drug reaction. In some cases, it can be associated with or caused by a viral infection, but it is not a true infection itself.

The treatment for OP is as follows: 1) remove the offending agent/drug, and/or 2) *treat with corticosteroids* (and use incredibly high "pulse doses" in severe cases). OP patients are notoriously in need of oxygen, often for prolonged periods and in high fractions, despite being fully conversant and not in overt respiratory distress. In most lung injuries, as the injury progresses, the air sacs in the lungs tend to fill with fluid, making the lung "heavier" and harder to inflate, a condition called acute respiratory distress syndrome (ARDS). For this reason, most patients with acute respiratory failure have to be placed on a ventilator.

Paul Marik, in his first four Covid patients, used a device to measure something called "extravascular lung water"; in all four patients, he found the lungs were *dry*. There was no excess water filling the lungs. Weird. We knew early on that Covid in the ICU, at least initially, was *not ARDS*.

Unsurprisingly, and in another sign of things to come, this critical observation was being ignored. So I knew that this was not ARDS initially and, in my mind, was almost certainly OP. Again, only a few others recognized this, like Cameron Kyle Sidell, another NYC intensivist whose video decrying this fact got almost a million views in the first surge.[1]

"Happy hypoxia" solved? Was SARS-CoV2 causing an organizing pneumonia . . . in almost everybody?

In order to validate my hypothesis, I called my friend and close colleague Dr. Jeff Kanne at the University of Wisconsin. Jeff was one of the world's premier chest radiologists and someone whose intelligence and skill in the field was and is nearly unparalleled.

"Jeff, what would you say if I told you that I think that all of these Covid patients are suffering from organizing pneumonia?" I asked. His answer? "Of course, they are. We wrote this up in March in the journal *Radiology* after an expert panel that I chaired completed our review of all the CT scans from Wuhan." They had actually written in their expert report that "the most common reported CT findings in Covid-19 patients *are typical of an organizing pneumonia* pattern of lung injury."

"Clinicians don't read radiology journals," I shouted into the phone. "We need to publish this in a clinical medical journal! Like NOW!" We quickly agreed that we would write it up together.

I went home after my ICU shift and started working furiously. The paper included radiographic, pathologic, and clinical evidence to try to prove that the pulmonary phase of Covid-19 was an organizing pneumonia and that the first line of therapy for this condition was (wait for it) . . . *corticosteroids*. I will say that my understanding of the cause of happy hypoxia has since deepened as I think another contributor to the severity of the hypoxia was the clotting and clumping of blood cells as they passed through the lungs, also affecting gas exchange. As masterful as my initial hypothesis was (I may be a tiny bit biased), it took four months and six journal submissions with two peer reviews before it was finally published in a prominent British journal.[2] The lowlight of that journey was the rejection letter I received after a rigorous peer review in the American journal *Chest*, when the peer reviewer who voted for rejection wrote, "In order for this paper to be published, a randomized controlled trial of corticosteroids would need to be performed."

I didn't know it at the time, but if my near future was going to have a theme song, those words would have made up the sad refrain.

CHAPTER THIRTEEN

Aerosol Transmission

Although the efficacy of masks is still somewhat controversial, my belief then, which was based on the best and only studies we had at the time, was that *the right kind* of mask, worn properly and in certain, specific situations for finite periods of time, could possibly be effective in preventing Covid . . . but not the flu. This is because transmission of Covid is markedly different from flu or other respiratory viruses, as Covid predominantly spreads via *aerosol transmission,* while most other viruses spread by large droplets. Understanding this difference is absolutely critical. In aerosol transmission, the virus is carried aloft by tiny floating droplets suspended in the air and then inhaled into the nose and throat and lungs by someone nearby. Large droplets, although they also travel through the air, will sink and land on surfaces and hands, and then can be transmitted by another who touches the droplets and then wipes their nose or mouth. It was obvious that aerosol transmission was occurring based on numerous reports of "super-spreader" events.

What I considered a major problem was the fact that both the CDC and the WHO were denying that aerosol transmission was occurring. So in April 2020, I started writing an op-ed to address this issue. The CDC would finally formally recognize the reality of aerosol transmission a full year later; it took the WHO two years.

I say this with humility, and because I am proud of the fact that I stood my ground in the face of so much adversity and criticism: "Lucky Pierre" struck again.

Here is the op-ed, published July 1, 2020, in *USA Today* and cowritten with Paul Mayo:

ICU doctors: Many more Americans need to wear N95 masks to slow Covid-19

Subhead: We believe that the lack of emphasis placed on the prevention of aerosol droplet transmission is a major contributor to the rising case numbers in many areas of our country.

Dr. Pierre Kory and Dr. Paul H. Mayo

Opinion contributors

As intensive care unit specialists in New York City, we are concerned about the increasing rates of Covid-19 infection, which may once again overwhelm our hospitals, in parts of the United States.

The severity, morbidity and mortality of Covid-19 must be re-emphasized to all, both young and old, as it spares neither. To avoid a catastrophic repeat of the initial surge, we recommend a population-wide intervention—a significant increase in the use of N95 masks—that might allow for a safer reopening of the US economy.

SARS-CoV-2 was thought to be primarily transmissible via large fluid-filled droplets generated by coughing or sneezing. Those droplets usually travel short distances before falling and will not reach another person practicing social distancing nor penetrate standard (cloth or surgical) masks.[1]

However, it has recently been determined[2] that a major mode of transmission of SARS-CoV-2 is via aerosol droplets, exhaled by presymptomatic, asymptomatic or symptomatic persons. These small aerosol particles remain airborne indoors for extended periods and can infect those nearby who inhale them into their lungs.

We believe that the lack of emphasis placed on the prevention of aerosol droplet transmission is a major contributor to the rising case numbers in many areas of our country.

What can be done to prevent the inhalation of these aerosols and therefore reduce the spread of SARS-CoV-2? The only mask that can prevent aerosol-size droplet inhalation is an N95.

Should there be widespread use of N95 masks by the general public? Not necessarily.

In Germany, Hong Kong, South Korea and Taiwan, Covid-19 was brought under control using standard masks (surgical or cloth, often homemade), an approach that relied on the fact that nearly 100 percent of citizens wore them in close-quartered public spaces.

Here is the key: Standard masks, although only partially effective in blocking inhalation of aerosols compared with the near perfect blocking performance of medical grade N95 masks used in hospitals, are highly effective at trapping the large droplets exhaled by infected people.[3]

These large particles downsize to aerosol size after emission when they undergo evaporative loss. Although standard masks are imperfect in both blocking or trapping, their combined performance when worn by both infected and noninfected persons leads to a low likelihood of transmission.[4]

The key point is that, for standard masks to be effective, there needs to be near universal wearing of these masks by all persons *when in any poorly ventilated, air-recirculated, confined indoor, or highly congested outdoor environment.* Conversely, emphasizing mask wearing in fresh air outdoor settings has no epidemiologic support and thus makes little sense.

Unfortunately, in some parts of the United States, the proportion of citizens routinely wearing standard masks in at-risk environments is nowhere near what's required to prevent spread. The fact that the maximal exhaled viral load of infected persons occurs before the development of actual symptoms should concern all who might come into close indoor proximity with maskless presymptomatic "super spreaders."

How should individuals protect themselves from infection in areas where near universal indoor mask use is not the norm? In such a situation, the best option is to wear an N95, which is designed to prevent inhalation of more than 99 percent of all droplets, large or small.

The United States has traditionally placed an emphasis on the rights of individuals. We respect the right of, but do not agree with, our fellow citizens who choose not to wear a mask. We also, however, are perplexed as to why responsible people would choose not to wear a mask given the potential harm, including death, that they could cause to their fellow citizens.

For those who live in an area where there is a high level of mask use, the risk of transmission or acquisition of Covid-19 infection is greatly reduced. For those who live in an area where there is a low level of standard mask use, the best approach for those who choose to wear a mask is to use an N95 mask to protect themselves from inhaling the exhaled aerosol droplets of their unmasked neighbors.

Those who protect themselves with an N95 mask would be able to safely participate in many activities involving groups of people. At-risk social gatherings and entertainment venues would be appropriate targets for routine N95 use, if universal standard mask wearing is not a precondition for entry.

The scientific and media publications describing "super-spreader events" provide some of the most damning evidence of the risks of congregating without universal mask wearing:

- The choir practice where one singer infected 52 of the other 60 attendees[5]? Aerosol transmission.

- The 22-year-old presymptomatic man who sang karaoke in an air-conditioned room for two hours and infected six of his 15 friends[6]? Aerosol transmission.
- The presymptomatic 29-year-old man who infected 102 people during visits to multiple crowded nightclubs?[7] Aerosol transmission.
- Cruise ship and aircraft carrier mini epidemics? Aerosol transmission through air-recirculation systems.
- The more than 100 meat-packing plants with massive outbreaks despite the workers' wearing safety goggles, gloves and gowns?[8] Aerosol transmission.

The challenge we face as a country is that we do not yet have sufficient N95 masks even for health-care workers, let alone for widespread distribution. It is unclear to us why this is the case, as it is well within the immense industrial capacity of the United States to ramp up national production of these simple low-tech safety devices.

Our current infection control strategies are unlikely to succeed. The public health approach will fail due to the imbalance between tracing and testing resources compared with the large and increasing number of documented and undocumented infections.

The "lockdown" approach employed by the Chinese government to rid Wuhan of the virus is not feasible in the open democracy of our country. Waiting for "herd immunity" would result in widespread death and disability while again overwhelming our hospitals.

While we wait for a vaccine, an effective cure, or the unlikely event that all 50 states will pass public health laws that mandate universal mask wearing in all indoor public places, could we have a national initiative to ramp up production of N95 masks so that all Americans have access to this level of protection?

Dr. Pierre Kory is a pulmonary and critical care medicine specialist who worked as an emergency volunteer caring for Covid-19 patients in New York City. He is also a founding member of the Front Line Covid-19 Critical Care Alliance that developed a Covid-19 treatment protocol. Dr. Paul H. Mayo is the academic director of critical care at Northwell Health LIJ/NSUH Medical Center and a professor of medicine at the Zucker School of Medicine of Hofstra University.[9]

In a bit of foreshadowing, this piece was initially accepted by the *New York Times*, but then dropped after the chief editor and my editor on the op-ed board were fired for publishing Senator Tom Cotton's editorial calling for the military to put down the Black Lives Matter protests. Despite my disappointment, I kept at it and soon after, I published it in *USA Today,* the newspaper with the largest circulation in the US.

CHAPTER FOURTEEN

The Trial of the Decade

For years I have conducted expert witness reviews of medical malpractice cases. The work is interesting and challenging and—I'll be honest—often well compensated. The lawyers I've worked with consistently rate my work highly; thus, I have been consulted often.

It was June of 2020, and I was mowing my lawn on a beautiful Saturday. I had just returned from that grueling month in NYC, and I was thrilled to be home and out of the grind for the first time since the Belfast conference in January.

My phone rang. I stopped the lawnmower and answered even though I didn't recognize the number—something I rarely did then and *never* do now. It turned out to be the agency that often contacts me to be an expert witness (via email only, though) to inquire whether I was available for either an ICU or pulmonary malpractice case. A phone call on a Saturday for a malpractice case? That was weird. The agent mentioned that it was a high-profile case that the lawyers had insisted would require strict confidentiality.

Would I be willing to maintain absolute confidentiality? Would I be able to produce a comprehensive report within ten days? Did I have any conflicts of interest with the Minneapolis Police Department?

Of course. George Floyd.

Protests were raging across the country. It would be the trial of the decade. But why were they calling *me*? His was not a medical malpractice case.

After confirming that I had no conflicts and promising strict confidentiality and a rapid review of the case, the agent confirmed that it was, in fact, Mr. Floyd's case. Within hours I was being interviewed by a team of lawyers on a Zoom call, and the next day they called to tell me that they had selected me to review the case as an expert medical witness.

I know. I had just been through a six-month bender of overwhelming work and stress, and I was jumping straight into what would likely turn out to be one of the trials of my lifetime. Sane people may wonder . . . *Why?* It turns out, I am cursed with an unshakable habit of saying yes to anyone who invites me to collaborate on a project. And this wasn't just any old project. How could I say no? As an expert in respiratory system physiology, I knew that it would be an interesting, challenging case. Plus, I was as upset and intrigued as anyone else following the story, and I wanted to understand what had happened.

The next ten days were somewhat of a blur as I spent nearly every waking hour producing my report. I read about the history of racist and violent behaviors by the Minneapolis Police Department, actions which had been carefully documented over the past hundred-plus years and included decades of incomprehensibly cruel and concerted actions against Blacks and others. I became familiar with the training on appropriate (and inappropriate) uses of force by police. I learned about the science and case histories of uses of "prone restraint," which is the use of manual force to hold or keep a person in a face-down position. It is a disturbing practice and considered a "never" event in policing. I felt physically ill just digesting this information, and this was before I started to watch the many videos collected from bystanders of the actual event.

The official expert testimony I submitted is a massive document which I'm honored to say my mentor Paul Mayo called, "a master class in pulmonary physiology." It's a difficult and long read, so I will summarize it in layman's terms here. To be clear, this was not a medical malpractice case; it was a homicide, as evidenced by the verdict reached by a jury of the officer's peers in the highly televised criminal trial and not from my report (I was the expert on the civil suit which determined the monetary judgment for the family). But George Floyd didn't die from (just) a knee on the neck. And it wasn't the drugs in his system. Just as with the response to the Covid pandemic, I think it is important to understand exactly what happened to Mr. Floyd so that it never happens again.

Before I go any further, a disclaimer: I recognize that the George Floyd case is a highly divisive and emotionally charged one for countless Americans. Many are profoundly saddened and outraged over the circumstances of his death; others are convinced he died of a drug overdose or believe he deserves little sympathy given his criminal background. Plenty are angered by the media coverage of the subsequent and oftentimes violent Black Lives Matter movement protests. So let me be clear: I am not a lawyer, a prosecutor, or a judge. I am not a political scientist, or a sociologist, or a psychologist. I was not asked to deliver a moral or character assessment of Mr. Floyd. (I have never met a saint, although I have known a few people that I feel come close.) I never met Mr. Floyd. I was simply asked for my expert opinion as a lung specialist as to the proximate cause of his death. It was an important task and one I felt confident I was expertly suited to perform.

You can read the full expert testimony on my Substack if you're curious.[1] For now, here is a brief, simplified summary of my conclusions:

Mr. Floyd was conscious and able to communicate for at least half of the duration of his ordeal, which suggests he did not stop breathing from an overdose of opiates. Drug overdoses occur within seconds to minutes of administration and the first physiologic effect is loss of consciousness. The opiate levels measured in his blood, although high, mean little given that chronic opiate users develop significant tolerance to such levels. Considering the prolonged time between when he would have had access to opiates and his death, it is nearly impossible to cite overdose as the proximate cause of death. I also do not believe that the pressure on Mr. Floyd's neck from Officer Chauvin's knee was the sole cause of his death by asphyxiation, as total occlusion of the trachea via external forces would have rendered communication impossible as well. My conclusion was that Mr. Floyd died by the combined forces of three officers bearing weight on his thoracic cage (upper back, lower back, and neck/throat) against a concrete surface, leaving him unable to take in a sufficient volume of air to expel enough carbon dioxide. Increasing levels of carbon dioxide in the blood first lead to unconsciousness, which is followed quickly by cardiac arrest.

Ultimately, it was my determination that Mr. Floyd was slowly and publicly suffocated as a result of prone restraint. It was a brutal, painful way to die and something I would not wish on my greatest enemy.

I watched hour after hour of eyewitness video footage while I was preparing my report, and the theme running through it all eerily mirrored what I was seeing all around me in Covid: the calculated, callous disregard for human life.

CHAPTER FIFTEEN

The FLCCC Takes Flight

During my frenzied work on non-FLCCC projects including my OP paper, my aerosol transmission op-ed, my resignation from UW, and the George Floyd case, the FLCCC was starting to spread its wings, especially in the wake of my (our) first senate testimony in May 2020.

When a post–Memorial Day weekend surge of Covid cases flooded Texas hospitals, media from all over the world came in to film in Joe Varon's United Memorial Medical Center in Houston. Joe humorously dubbed himself "the Covid Hunter," as he was drowning in Covid patients and hungry media. Joe's efforts in Houston were massively heroic. He literally worked in the ICU for 715 days straight until he took a break to attend his daughter's wedding. Joe tallied his media appearances throughout because they were so overwhelming. He ended up averaging seven TV and radio interviews a day during the peak of the pandemic with a maximum of eleven in one day. Overall, since March 19, 2020, he has done 3,428 media interviews: 60 percent in English, 38 percent in Spanish, and 2 percent in something other than those two languages from a total of sixty-seven different countries. One highlight of all of this media attention was a moving picture of Joe hugging one of his patients that went viral and was shared around the world.[1]

Local Houston media, Sky News, the BBC, the *Los Angeles Times,* and CNN all interviewed him, but unfortunately the focus of most of the reports was on the surge of cases rather than the MATH+ treatment protocol. This might have been when the phrase *Clown World* made its way into my daily vocabulary.

By June 2020, our MATH+ protocol was routinely being shared by several allied health-focused educators, organizations, and institutions, but the legacy media remained decidedly uninterested. Slowly, we began receiving modest but welcome coverage in the UK (Sky News), on a handful of regional US television stations, and in a few local newspapers. In the meantime, Covid continued to rage across America, leaving a path of destruction and death in its wake.

With a few rare and notable exceptions, media avoidance of our protocol(s) continues to this day. No major media has published or aired a story on our MATH+ Hospital Treatment Protocol for Covid. Even after we announced our extremely low rate of mortality—less than 6.1 percent compared to 15 percent at best and in many cases far, far higher in hospitals around the country and world; even after treating nearly 450 patients with MATH+ within six hours of presentation to Paul's and Joe's hospitals; even though we had some of the best and brightest in their fields on our team; they ignored us.

Paul and Joe found that the few patients they treated who did not survive had either succumbed to comorbidities or had presented in an advanced hyper-inflammatory state. My personal experience with MATH+ was quite different. There was a relatively simple reason for that: In both Joe's and Paul's hospitals, they started MATH+ within hours of arrival. They were able to do this because they ran the clinical services in their respective hospitals and their protocol was followed by those who worked under them. In the hospitals I would work in over the next year and a half, I was not in a clinical leadership position. By the time the patients came under my care in the ICU, they had been undertreated or not treated for many days in the regular hospital wards. Starting MATH+ late in their disease course was much less effective. I tried to convince the hospitalists on the wards to start some of the combination of treatments before the need for ICU but was only occasionally successful.

Finally, in July 2020, an Oxford University Randomized Evaluation of Covid-19 Therapy (RECOVERY) trial of dexamethasone reported a significant mortality reduction in Covid patients requiring supplemental oxygen or mechanical ventilation. This was essentially the corticosteroid treatment I had testified about in Senator Ron Johnson's Homeland Security Committee hearings. Corticosteroid treatment became the standard of care overnight, although sadly this still didn't result in any notable recognition of our protocol—of which corticosteroids were a mainstay. I received a memorable text from one of my favorite mentees, Joe Mathew, the morning the paper was made public. "We should have listened to Pierre," he wrote. Indeed.

One of the pivotal events which led to the evolution of the FLCCC into a formidable nonprofit organization was when a man named Lenn Getz, the head of the New York Cancer Resource Alliance, heard about our group and contacted me. Lenn offered to help spread the word about MATH+ to the many physicians who supported their alliance. He was a highly experienced leader of nonprofit health organizations, and in our many conversations over months, it was Lenn's guidance and support that got me to shell out even more money on lawyers who helped set up the FLCCC as a 501(c)(3) nonprofit organization. Prior to that point, the public face of the FLCCC was what I called "five guys with a website," and we had no personal liability protection if anyone were to be harmed by our advice and/or decided to take legal action against us. Not only did incorporating our group as a nonprofit organization protect us individually, but it also created a framework under which we could accept donations, something that I was very hopeful for given the amount that was riding on my credit card. Keep in mind that I was not working at the time, so our five-person, formerly double income household was now operating on a single salary, all while I continued to throw money at the FLCCC to keep it afloat.

Prior to my resignation from UW, I had tentatively secured a new position at a major medical center in Milwaukee, a little more than an hour from my house. Although a large, sophisticated facility, it was not a teaching hospital. My career as a medical educator was over. But the money was good, and if I wanted to overwork myself, my earnings would be more than I had ever made in my career.

As it can be inclined to do, the universe took a little piss on my parade. I discovered that the credentialing process would take five months.

Desperate to find work in the interim, I reached out to a few staffing agencies to see if I could find work as an independent contractor. Hospitals around the country were desperate for specialists such as myself to run their ICUs, and I accepted a position at a teaching hospital in Greenville, South Carolina. It was a three-month gig and I would be there two weeks a month, living in a hotel, but I would have an income again. That left two months to focus on FLCCC and Covid stuff before starting my bi-monthly commutes to South Carolina.

In the space of those two months, I worked on multiple projects related to the FLCCC. The first one was writing a paper marshaling the evidence for all the individual components of our protocol, entitled, "Clinical and

Scientific Rationale for the Covid-19 MATH+ Protocol." I foolishly hoped that, with publication of the evidence supporting our protocol, it might get more attention by the outside world. (Spoiler alert: It was not an RCT, so it was roundly ignored). It was a beast of a paper with over two hundred references and somewhere in the neighborhood of 12,000 words—an almost unheard-of length for a scientific manuscript. Since each component had tons of mechanistic and clinical supporting evidence and we had almost a dozen therapeutic components in the protocol, it turned into a medical *War and Peace*. It also didn't help that my reference manager software sucked, which meant I had to manually reorder two hundred references over and over. Try doing that for hours a day and keeping your sanity.

One day, the FLCCC got an email out of the blue from a graphic designer named Frank Benno Junghanns[2] in Berlin. Frank not-so-gently informed us that our website design was amateurish and that the entire site as well as our printed protocols lacked a clean, scientific, professional look.

He wasn't wrong, but it wasn't as if any of us had graphic designer on our resumes.

Frank's suggestions for improvements were many and impressive, and he generously offered a greatly reduced rate for his work. I might have sprained a finger in my haste to reply with an official job offer.

Frank was a magical, much-needed addition to the team. Tireless in translating the protocols into the most common languages, he also redesigned the FLCCC logo and website to appeal more directly to the medical community. He later came up with the idea of creating a broader basis for the dissemination of the MATH+ protocol by reframing our name from a "Working Group" to an "Alliance." With that, the Front Line Covid-19 Critical Care Alliance was born, and eventually grew into the name almost everyone uses now, which is the FLCCC. Translations of the MATH+ protocol were posted on our website in six languages (today it is dozens) and several of us gave online talks explaining the protocol to doctors in India, Bolivia, and Argentina.

The FLCCC urged doctors who were in support of the MATH+ protocol to join the Alliance. By August, there were thirty names posted on our newly redesigned website. Despite mammoth battles with my reference manager, I finally submitted the extensive scientific review of the pathophysiologic and clinical evidence supporting the use of each medicine in MATH+, written over the preceding months by myself and the other members of the FLCCC.

Tens of thousands of people were viewing FLCCC's posts on social media, and many were reaching out, desperate to know what hospitals they could go to for Covid treatment to be assured of getting the MATH+ protocol. There were a scant handful of options: Joe's hospital in Houston, Paul's in Norfolk, Virginia, and—if you could find them—the odd rogue intensivist in a random hospital here or there who was quietly employing the protocol.

MATH+ was turning around even the sickest patients. One dramatic case involved a patient named Dr. Manny Espinoza, himself a urologist. Manny was rapidly deteriorating in another hospital, and his wife, Dr. Erica Espinoza, a family practice physician, demanded he be transferred to a MATH+ facility. Six air ambulance companies refused to transport him because of his Covid positive status. Finally, Erica found an ambulance who would take her gravely ill husband to United Memorial Medical Center in Houston. Once introduced to MATH+, Manny went from death's door to nearly fully recovered within days. An emotionally moving video of the Espinozas' powerful story is one I share often.[3] If you can watch it without tearing up, you're a stronger person than I am.

Although people were now following the FLCCC in droves, the medical establishment and media continued to ignore our group's collective expertise, rationale, and early treatment successes. They refused to report on the protocol because we had not conducted a randomized controlled trial. Meanwhile, several accomplished educators and physicians were taking notice. One of them was Dr. Mobeen Syed, better known as "Dr. Been" to hundreds of thousands of medical professionals and students in 182 countries who follow his instructional videos on Facebook and YouTube.[4] Dr. Been is both a physician and a software engineer and has used his unique skillset to pioneer several innovative medical devices, including a portable 3D ultrasound system.

"I believe that MATH+ with aggressive early intervention is the most comprehensive and the best choice available to the medical community," Dr. Been said during one of his four separate video lectures on the MATH+ treatment protocol that he uploaded to his channels. "It's the most important core management approach to save thousands of lives. If it were up to me, I would make MATH+ a mandatory protocol for Covid-19 management."

Across the pond, at least, folks were paying attention. A prominent group of ear, nose, and throat and ICU experts from the Ukraine published a paper recommending MATH+ be used in hospitals there. Paul was later told by one of the doctors that it had become a standard protocol in Ukrainian hospitals. If only the US had jumped on board so quickly.

At that time, over six million cases of Covid had been confirmed in the US and more than 180,000 Americans had lost their lives to the virus (allegedly; it would be months before we'd know about the faulty PCR tests and inflated case counts of people dying "with" Covid and not "from Covid"); both figures represented the most of any country in the world. Health officials were warning that thousands more deaths were likely in the next few months. Finally, in December 2020, our MATH+ passed peer review and was published in the *Journal of Intensive Care Medicine* . . . to a delightful chorus of crickets.

Worse, that paper would later be attacked and retracted, before we ultimately would revise and republish it in the *Journal of Clinical Medicine Research*.[5]

Even before the retraction, the trade newsletter MedPageToday, which is often populated with opinion pieces planted by Big Pharma PR agencies, published a "news" story headlined, "Doctors Publish Paper on COVID-19 Protocol; Experts Unconvinced."[6] Never mind that we *were* experts; apparently, other, *better* experts needed more evidence. And by "more evidence," of course, they meant randomized controlled trials. We were essentially accused of "throwing spaghetti at walls and hoping something would stick."

The first six months of the FLCCC certainly resulted in better outcomes for many patients, but little did we know that we were teetering on the brink of a revolution. Paul was about to identify ivermectin, an inexpensive, incredibly safe, generic, repurposed drug as an immensely effective and potent therapy against SARS-CoV2. It was a discovery that could and should have saved lives and ended the pandemic—if not for one major problem: Repurposed drugs like ivermectin are generally off-patent, which means the manufacturer has lost exclusive marketing rights. In other words, competitors can make and sell dirt-cheap versions. So not only is repurposing of little interest to the original manufacturer (in this case, Merck), but also to any other Big Pharma companies trying to develop or promote pricey alternatives.

What should have been met with a life-size gold statue or at least a ticker-tape parade instead spawned a period of censorship, destruction, and awakening that would change not only me and my colleagues in the FLCCC but basically the entire world in almost every conceivable way: professionally, intellectually, politically, financially, physically, and even emotionally.

Until this point, we were a volunteer army fighting on the front lines of the battle against Covid. We were about to lead a war.

PART TWO

THE WAR ON IVERMECTIN

CHAPTER SIXTEEN

A Potential Solution to the Pandemic

.

No one has, as far as I know, written the book, *The War on Hydroxychloroquine*, but they certainly could have and probably should have. From the moment President Trump first spoke about a promising drug called hydroxychloroquine (HCQ), the defamation campaign was swift and furious.

"HYDROXYCHLOROQUINE & AZITHROMYCIN, taken together, have a real chance to be one of the biggest game changers in the history of medicine," Trump tweeted, citing the *International Journal of Antimicrobial Agents* as his source.

The top Twitter reply, ironically from a man whose bio reads, "do good and it gets good," reads like this:

"Why? Is the Republican Party allowing your treacherous corrupt MADNESS! Your moronic INCOHERENT BULLSHIT RHETORIC is dangerous! You delusional treasonous pathological liar fucking moron conman rancid fishlips incompetent corrupt clueless ASSHOLE! Stop this misinformation! LIAR!"

You know a guy's got something really important to say when he uses more all-caps and exclamation points than Trump.

By the summer of 2020, HCQ was trending. A group of physicians calling themselves America's Frontline Doctors (AFLDS) held a press conference on the steps of the United States Supreme Court in Washington, DC. Led by Dr. Simone Gold, a tireless medical freedom fighter, the "White Coat Summit" warned of the ineffectiveness of masks, social distancing, and lockdowns, and spoke of the promise of repurposed medications—in particular, HCQ—in early treatment.

The recorded conference went viral, thanks in no small part to the Trump family's enthusiastic sharing on social media. It racked up more than 14 million views before the "disinformation police" could stymie the spread. While streaming platforms scrambled to remove the video, media spin doctors were working overtime to slander not only what the AFLDS was recommending but attacking the group as a whole (routinely referring to it as a "right-wing political organization") and individual members in turn. They were ridiculed by reporters and accused of—wait for it—*spreading deadly misinformation* by academia. I had never seen anything like it.

The medical community had been holding its collective breath waiting for what would be considered large-enough, *proper* RCTs of a number of interventions that were being used from early on in the pandemic, including HCQ, convalescent plasma, the cytokine blocker tocilizumab, and the antiviral interferon. In September of 2020, the results of studies on these began to be posted or published. One after another, each was found "ineffective." Although this would be an entirely predictable fraud to me now, back then, Paul and I believed in the integrity of the trialists and the prestigious journals they were published in. We were still naive to the fact they were completely manipulated by researchers working for Big Pharma with the intent to destroy the evidence of efficacy of HCQ in particular.

Enter Dr. David Boulware, an infectious disease specialist from the University of Minnesota. Boulware conducted one of the first RCTs on HCQ, which he published in the *NEJM* in August of 2020.[1] His was one of the only two substantive papers used by the FDA to justify withdrawing the EUA for HCQ. Since the other paper was the UK RECOVERY study in hospitalized patients which used near-lethal doses (a violation of research ethics and a crime against humanity which still needs to be addressed), Boulware's paper was the only one addressing post-exposure prophylaxis. Accordingly, the importance of this paper's impact on the pandemic cannot be overstated, both in terms of saving lives and in shaping how repurposed drugs would become marginalized.

A formidable warrior in the now over three-year fight against research fraud is David Wiseman. David is a research bioscientist with a background in pharmacy, pharmacology, immunology, and experimental pathology and one of the top sixty-six research scientists at Johnson & Johnson, where he headed up a comprehensive research and development program focusing on surgical adhesions. Since 1996, he has run his own R&D drug, device, and biologic consulting business and cofounded the world's first clinic for the integrated treatment of pelvic and abdominal pain and pioneered the use of a device for those conditions. In Covid he came out fast and furious. His raw dataset analyses have overturned previously negative studies used to justify public policy regarding hydroxychloroquine and ivermectin. He has made over twenty-six submissions on Covid vaccine safety and efficacy to government bodies including the FDA, CDC, and NIH. Additionally, he is a contributor to *TrialSiteNews* and participated in Senator Johnson's two expert panels on Covid in 2022.

David reached out to me after he found a statistically significant correlation between time to treatment and efficacy in HCQ, which he sent to Boulware. When he received no response, he penned a letter to the *NEJM*, which I coauthored along with Dr. Mayur Ramesh, an infectious disease specialist affiliated with Henry Ford Hospital in Detroit. The letter was rejected, but within two minutes of receiving the now-released raw data, it was obvious that a medical blunder with massive implications had occurred. The Boulware paper failed to account for the shipping time of the drug to study subjects. This is critically important if the goal is for HCQ to be a sort of "morning after" pill.

Wiseman carefully documented this in a re-analysis protocol posted on medRxiv, an online archive for unpublished medical manuscripts. Weeks and dozens of emails later, Wiseman persuaded Boulware to retrieve and provide the FedEx shipping time data. The first set of shipping data was inadequate. A revised set shifted Boulware's published 17 percent estimate of HCQ efficacy to a potentially pandemic-changing, statistically significant 42 percent, if the drug was delivered up to three days after exposure.

Wiseman, like me, believed that this gaffe was the result of miscommunication between Boulware, his statisticians and pharmacists. The study was otherwise well-designed, albeit launched quickly due to pandemic conditions. Under the circumstances, the UMN team could have been forgiven for this innocent, albeit significant error, if it had been acknowledged and corrected as soon as it was discovered.

Shockingly, Boulware instead responded by accusing Wiseman of falsifying data and a flurry of emails ensued that included (at Wiseman's instigation) UMN's vice president responsible for research misconduct policy. Backing down but without apology, Boulware confessed that he had been unaware of how his database had defined certain time periods or events. Boulware's ignorance of his own data, also reflected in information he had provided to Wiseman earlier, was the source of Boulware's erroneous accusation against Wiseman, which seemed like an attempt at misdirection away from the fundamental study blunder. To date, he has not corrected his paper within the pages of the *NEJM*.

Wiseman also found major issues in two other papers from Boulware's group (Skipper et al. and Rajasingham et al.). He repeatedly asked for the raw data from the authors and the journals they published in but was denied by all.

At the time I was working with Wiseman on this, I simply thought Boulware was someone of questionable ethics and honesty; a careless, biased schmuck. Wiseman, on the other hand, already suspected Boulware was a cog in a much larger, more malevolent machine. Our paper[2] exposed the misrepresentations and manipulations Boulware had pulled in his HCQ trial. The *NEJM* rejected our paper (which fully explained Boulware's "confessed" error) within eighteen hours and without peer review because it lacked "focus, content and interest."

We were not surprised by this. But we wanted our concerns and objections documented, on record forever.

There was no way to know it at the time, but Boulware would go on to serve as a major enemy asset in the later war on ivermectin. By the end of 2022, Boulware was one of the main "go-to" media commentators on ivermectin given he was a coauthor on multiple large ivermectin studies, even though they had design and conduct deficiencies that would make his HCQ trial shenanigans pale in comparison.

Anyway, as one "big, important RCT" after another purportedly "proved" HCQ to be ineffective, we bought the lie, at least for a while; HCQ didn't work.

At that time, the FLCCC did not yet have an official early treatment protocol for Covid. (Paul's original protocol still lived on the EVMS website, however.) Thanks to fraudulent data and corrupt "science," it would be a long time before we learned the truth about HCQ and included it in our recommendations.

On an FLCCC Zoom call one day in mid-October of 2020, Paul excitedly spoke about a possible solution to the pandemic. A dirt cheap, widely available, Nobel prize–winning antiparasitic medication called ivermectin was showing early but great promise. At the time, the "evidence base" Paul was making this assessment on was quite small and consisted of maybe a handful of studies posted on preprint servers, plus one brand new, peer-reviewed trial published in the highly regarded journal *Chest* by Juliana Rajter et al., the wife of the senior author Dr. Jean-Jacques ("JJ") Rajter.

The results of that single study were staggering as they reported a massive reduction in death with ivermectin treatment.

JJ and Juliana were both pulmonary and critical care specialists and, like us, were trying to identify an agent that would work. They knew of reports showing ivermectin was effective in vitro (studied outside of a living organism) against Zika and HIV. When an in vitro study published in early April of 2020 showed that ivermectin completely inhibited the replication of SARS-CoV2, they decided to try it with their patients. They began documenting their experiences on Facebook. On April 12, 2020, Juliana posted:

> I feel I need to share this with the public at large. I have used ivermectin now on multiple carefully selected patients with a 100 percent response rate within 48 hours. Avoided intubation at all costs even if requiring 100 percent FiO2. All of them have improved. The earliest patients in the series are now on 2 liters/minute nasal cannula and afebrile. No side effects noted. In my opinion as a pulmonary specialist, this is amazingly effective and should be considered in carefully selected patients. I am trying desperately to get the word out, but as a community pulmonologist who does not work for [an] academic institution, this has proven extremely difficult.

By June of that year, the Rajters had treated enough patients to conduct a study comparing those treated with ivermectin versus those who had not been given the antiparasitic drug. They identified a control group whose characteristics matched as closely as possible the characteristics of the treated group in terms of age, sex, and severity of illness, something called "propensity matching," a highly valued statistical and study technique with results that, on average, match those of RCTs.[3] What they found was, to use Paul's signature phrase, "truly astonishing": a 47 percent reduction in death among all treated patients, and an incredible 73 percent reduction in death among the most severely ill. If you recall from earlier in this book, the "number needed to treat" or NNT is a measure of a therapeutic's potency. According to this study, ivermectin's NNT worked out to be . . . *two*. That means for every two severely ill patients you treat with ivermectin, one patient who otherwise would have died would be saved. It was unprecedented.

Pandemic over, right?

Of course, you know the tragic answer to that. But at the time, we believed that we'd discovered the Covid holy grail. Inspired by what I had learned thus far, I began my own exploration into the available ivermectin research. I was shocked to discover that since 2012, over a dozen in-vitro studies had been published showing that ivermectin stopped replication of more than ten different viruses including Zika, West Nile, HIV, and influenza. Almost all were RNA viruses . . . just like SARS-CoV2.

Ivermectin was not only a broad antiparasitic but also a broad antiviral? This wasn't huge; it was monumental. No, it was absolutely *epic*.

Paul uploaded a lecture to YouTube titled "Covid-19: Saving the Planet with Ivermectin and Masks." It's since been removed, of course, and obviously we didn't get the mask part right, but the ivermectin information was spot on. As more and more studies started popping up on preprint servers from centers across the world, I was like a kid in a candy shop with his daddy's credit card. I proposed to Paul that I write a review paper compiling all the data from the studies we were finding.

In my research I came across another incredible preprint paper[4] linked from *TrialSiteNews*, an online research and pharmaceutical industry news site. *TrialSiteNews* was and is an absolute treasure trove of life-saving information, and their reporting on emerging ivermectin (and vaccine) data was unparalleled. They were doing something that no one else was doing: showing data from both sides of the story, objectively, and without conflict of interest. *TrialSiteNews*'s CEO Daniel O'Connor, an incredible man turned close friend, was one of those who knew early on that the world had gone mad and was trying to get real-time, objective, life-saving information to the public. Although Big Pharma read the site, he had other "customers" as well: everyday people. Daniel was a truth-teller, and he took his responsibility to *all* of his readers seriously.

The paper I found there showed the jaw-dropping impacts of a mass ivermectin distribution program in Peru called Operation Tayta. I was trembling as I studied the graphs.

The analyses showed that in every region where ivermectin had been distributed, cases and deaths peaked and then rapidly fell, all with the same temporal association which was eleven days after each mass distribution started (see Figure 3, opposite page). Further, in the one place where ivermectin had not been distributed, Peru's capital of Lima (the background columns that resemble massive mountain ranges in the Figure 5 graph on page 101), cases and deaths continued to rage.

Paul was right. *Widespread ivermectin distribution could end the entire pandemic.*

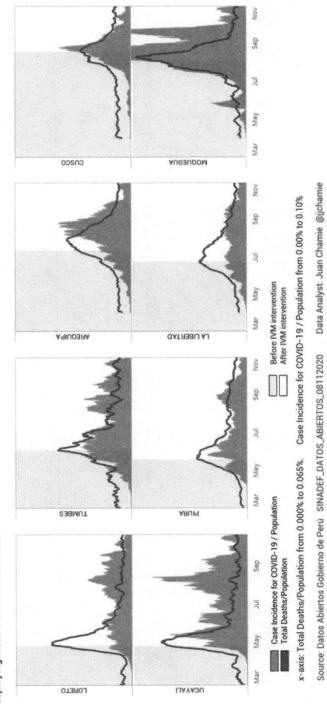

Figure 12. Total Deaths/Population and Case Incidence for COVID-19 / Population in population older than 60 years old for eight Peruvian states deploying mass ivermectin treatment

Case Incidence for COVID-19 / Population
Total Deaths/Population

Before IVM intervention
After IVM intervention

x-axis: Total Deaths/Population from 0.000% to 0.065%.   Case Incidence for COVID-19 / Population from 0.000% to 0.10%
Total Deaths/Population from 0.000% to 0.065%.

Source: Datos Abiertos Gobierno de Perú   SINADEF_DATOS_ABIERTOS_08112020   Data Analyst: Juan Chamie @jjchamie

Figure 3. Excess mortality (line) and Covid-19 cases (shaded) before and after ivermectin distribution began at different times in 8 departments of Peru (shaded on left). (Source: Datos Abiertos Gobiernos de Perú – datosabiertos.gob.pe)

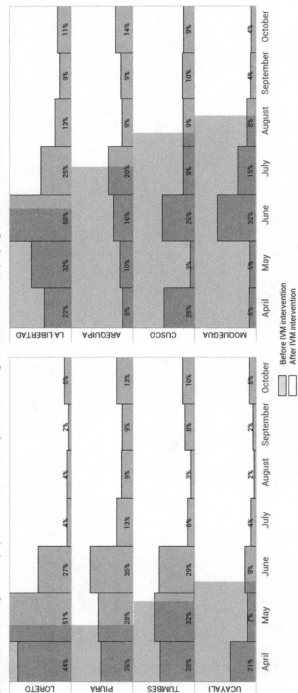

Figure 13. Case Fatality Rate in population older than 60 years old for eight Peruvian states deploying mass ivermectin treatment

Source: Datos Abiertos Gobierno de Perú SINADEF_DATOS_ABIERTOS_08112020   Data Analyst: Juan Chamie @jchamie

Figure 4. Infection fatality rates (bars) before and after ivermectin distribution (shaded on left) began at different times in 8 different departments of Peru. (Source: Datos Abiertos Gobiernos de Perú – datosabiertos.gob.pe)

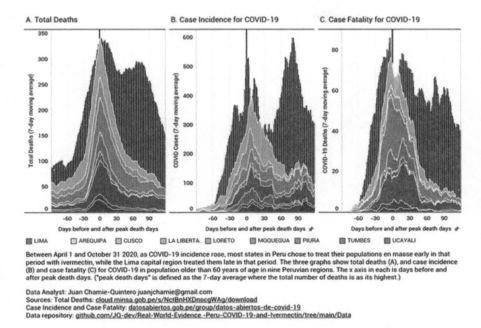

A. Total Deaths   B. Case Incidence for COVID-19   C. Case Fatality for COVID-19

■ LIMA   □ AREQUIPA   ▣ CUSCO   ▣ LA LIBERTA.   ▣ LORETO   ▣ MOQUEGUA   ▣ PIURA   ▣ TUMBES   ▣ UCAYALI

Between April 1 and October 31 2020, as COVID-19 incidence rose, most states in Peru chose to treat their populations en masse early in that period with ivermectin, while the Lima capital region treated them late in that period. The three graphs show total deaths (A), and case incidence (B) and case fatality (C) for COVID-19 in population older than 60 years of age in nine Peruvian regions. The x axis in each is days before and after peak death days. ("peak death days" is defined as the 7-day average where the total number of deaths is as its highest.)

Data Analyst: Juan Chamie-Quintero juanjchamie@gmail.com
Sources: Total Deaths: cloud.minsa.gob.pe/s/NctBnHXDnocgWAg/download
Case Incidence and Case Fatality: datosabiertos.gob.pe/group/datos-abiertos-de-covid-19
Data repository: github.com/JQ-dev/Real-World-Evidence-Peru-COVID-19-and-Ivermectin/tree/main/Data

Figure 5. Deaths (A), Case Incidence (B), and Case Fatality Rates (C) of Covid-19 before and after the peak impact within 8 departments of Peru with ivermectin distribution programs (center lines) compared to the capitol of Peru (Lima) which did not distribute ivermectin (shaded in background). (Source: Datos Abiertos Gobiernos de Perú – datosabiertos.gob.pe)

The paper's author was Juan Chamie, and his background was fascinating. He was an independent data analyst in Cambridge, Massachusetts, who routinely worked for large, well-known corporate clients such as Wayfair and Trivago. A native Colombian, Juan had heard about the efficacy of ivermectin from friends in South America early in the pandemic, so he began analyzing extensive data from public health records around the globe.

I read the paper twice and although it needed some work (Juan is a genius, albeit not experienced in writing academic papers), the conclusions were incredibly well-supported by the data. I couldn't wait to call Paul.

"Are you near a computer?" I asked.

"Yes," Paul replied.

"Check your email, read this paper I'm sending you and call me back. Immediately. Ivermectin can end this thing."

Juan's paper was the final missing puzzle piece allowing me, Paul, and the FLCCC to conclude that ivermectin should be globally and systematically

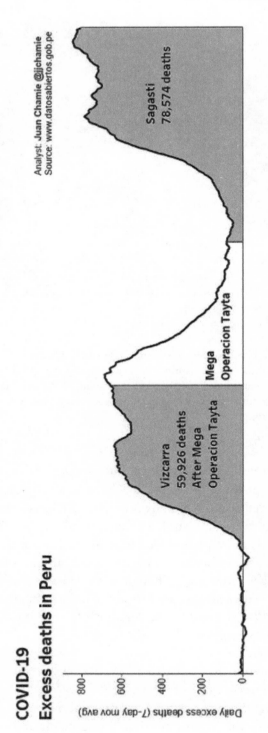

Figure 6. Excess deaths in Peru before and after Mega Operacion Tayta. (Source: Datos Abiertos Gobiernos de Perú – datosabiertos.gob.pe)

deployed in the prevention and treatment of Covid-19. I recruited Juan to work for the FLCCC, where his contributions were profound. Just as they were on Twitter, where he got de-platformed more times than I can count for posting, over the next eighteen months, relentless epidemiologic analyses and striking graphics showing Covid cases and deaths dropping off the cliff in countless regions, cities, and countries after the introduction of ivermectin.

In a bit of foreshadowing, what happened in Lima versus the rest of Peru was a pattern that would play out over and over again around the world in regard to physician willingness to consider using ivermectin. City doctors versus rural doctors. Red versus blue doctors. "System employed" doctors versus private practice doctors. Rich countries versus poor countries. The big city academics and centers all dismissed and derided ivermectin for having "*insufficient evidence.*"

Unfortunately, and to that end, Peruvian President Martin Vizcarra, who had approved Operation Tayta in Peru, lost his bid for reelection that November. His replacement Francisco Sagasti disbanded the program. Cases and deaths subsequently skyrocketed (see Figure 6, left).

In mid-October of 2020, the team and I began compiling all of our research. I embarked on a four-week frenzy of work unparalleled in my career—which is saying a lot, as I have suffered from extreme workaholism since before becoming a doctor. At that time, I had just started a new, permanent (yeah, right) position, working full-time shifts in a major medical center in Milwaukee. Less than a month later, on November 13, 2020, I uploaded the first draft of our comprehensive narrative review paper on ivermectin, which included wildly positive summary analyses of both the prevention and treatment trials along with Chamie's epidemiologic analyses of ivermectin. That paper still sits on the very same preprint server today.

Simultaneously with the paper, we created our first early treatment protocol (separate from our hospital protocol) which we christened "I-MASK+" (later changed to "I-CARE") and included an initially cautious dose of ivermectin as we were still learning and analyzing the data. We quickly and continuously increased the dosing, frequency, and duration according to the severity and viral loads of the subsequent variants (while incidentally, all the "high quality" Big Pharma/NIH trials did the opposite).

The FLCCC's ivermectin journey had begun.

CHAPTER SEVENTEEN

Angels Fall from the Sky

Although our protocol was getting attention, cases and hospitalizations were rising exponentially across the country. The FLCCC was like the angry mom at the back of the auditorium at a heated school board meeting: we had plenty to say—we just needed a microphone.

As our intrepid communications director and creative director respectively, Joyce and Betsy were determined to get the FLCCC noticed by the media. People were dying, we had a solution, and the vast majority of the world was simply sitting back and allowing the carnage to continue. It was madness. Joyce brilliantly suggested that we hold a press conference to announce our protocol. It was what AFLDS had attempted to do with hydroxychloroquine, and we knew how *that* had gone down, but by this point we had mountains of science on our side. Surely, we wouldn't suffer a similar fate.

Cue the laugh track.

In the weeks leading up to the press conference, a man named Sean Burke had reached out to me on multiple occasions to brainstorm ideas on how to get word of ivermectin to the masses. Sean had an interesting background; he was a writer, radio show host, entrepreneur, and philanthropist. He had worked for some of the biggest names in the entertainment industry, founded numerous companies including a nonprofit which got four privacy bills signed into law over a five-year period in California, appeared on numerous TV programs, and been quoted in all the major newspapers. In 2016, Sean bought a giant RV and had it painted in patriotic red-white-and-blue and emblazoned with "Let's Press America's Reset Button" across the side. He then embarked on a national tour talking to Americans and media outlets in an effort to raise awareness of our ballooning national debt. We could use about 100,000 Sean Burkes right now.

Sean was friends with Cass Warner (of Warner Brothers fame), whose medical care Paul had serendipitously and successfully consulted on while she was deteriorating from sepsis in a Dublin hospital years earlier. Paul's interventions and control of her hapless ICU doctors saved her life. After learning about our planned press conference, Cass gave Sean a generous grant to shoot a documentary[1] about the FLCCC and our MATH+ hospital protocol (which we rebranded I-MATH+ when we added ivermectin as we had done with our outpatient protocol). That documentary, *What is Ivermectin?* details the discovery and development of this wonder drug and highlights its status as one of the most important public health advancements in history. Sean did a stellar job, and with the guidance of accomplished director Adrian Urso, it became part of our growing press kit.

Paul, Joe, and I headed to Houston for our press conference, which you can watch on the FLCCC website.[2] If you do, you'll notice something comical in the opening scene: Joe is speaking, and Paul and I are on either side of him. Wearing masks. Outside. (This was another classic Clown World move, similar to the entire planet being forced to wear masks while walking to our tables in restaurants but magically disallowing them as soon as we were seated.) I get crushing chest pain watching it now, but we would have gotten murdered by the press at the time had we dared to show our faces.

That press conference focused on the prevention and early treatment of Covid and it made a decent ripple, scoring coverage on several local Houston TV stations, across Central and South America on Spanish language media stations, and even a couple of outlets in Europe. The Associated Press was there too. Many supporters later reached out to us to say they had watched it live in Europe and across the nation in cities like Chicago and New York.

We were finally gaining traction. Suddenly, several critical people appeared in the FLCCC's orbit, all of whom would play vital roles in our development and success. One was a wealthy businessman named Ronny Scerbar (not his real name). Ronny sent us a generous $2,500 donation with a handwritten note saying that he would like to support a national, public-service-announcement type TV commercial campaign. I was struck by the boldness of this idea and reached out to him immediately. We became fast friends.

Our first conversation lasted for hours and was absolutely fascinating. It turned out that Ronny had learned of ivermectin's potent efficacy back in March of 2020 from his many contacts in Iquitos, Peru. They had told him that *no one taking ivermectin was getting sick or dying,* and he had been taking it weekly as a preventative ever since. In fact, he had been sharing the evidence of efficacy with his entire company of 300 people, and about seventy-five of his employees had been taking it regularly as well. *Not one of them had contracted Covid.* He was a Trump supporter and told me that through a close contact he was able to send the president a one-page memo summarizing the mechanisms and clinical efficacy of ivermectin against Covid. The day Trump's own Covid diagnosis was announced, Ronny's contact flew to Washington and gave the memo to Paula White, Trump's spiritual advisor. Word came back to Ronny that Trump had subsequently received ivermectin as part of his treatment. He was out of the hospital two days later.

I'll add here that this was before the FLCCC included ivermectin in any protocol; we had nothing to do with it. I just find it interesting that this information was never reported publicly (although I suppose not surprising).

During our initial conversation, Ronny pledged $25,000 to our organization. When I received the check two days later, it was not for $25,000. *It was for $250,000.* An anonymous angel saved the day.

It was around this time that we learned of Chris Martenson of PeakProsperity.com fame. Chris was a brilliant, bestselling author with a popular podcast and a practical perspective who had been posting daily Covid content long before nearly everyone else on the planet—until the Censorship Police found him, of course. He, too, was outspoken about the promise of ivermectin, so we arranged to meet in Texas. As with so many of my new like-minded colleagues, we became fast friends; later, the FLCCC would invite him onto our board. Chris is an enormously respected scholar and observer of the world around him, and he lends expert guidance not only to the organization but to me personally. I am humbled by his friendship.

Another early angel who was critical to launching the FLCCC was a wealthy businessman named Jeff Hanson. Jeff had been a wunderkind in the health-care real estate investment trust (REIT) industry whose investment fund was so successful that it is now about to go public—a rarity in the REIT world. After seeing my testimony, Jeff contacted Paul the next day and they had a long chat. He followed up with a donation of $100,000—the first of several even larger checks he would write to the FLCCC. Jeff wasn't just instrumental in our growth; he's also a warm and hilarious human who has become a dear friend.

In one of my earliest conversations with Jeff, Chris Martenson's name came up. Jeff had been following Chris since early 2020 and had been (wisely) making decisions throughout the pandemic based on Chris's predictions and recommendations. Jeff and Chris connected during a second trip we all made to Joe's hospital to shoot more footage for the documentary. The two have since become close friends and serve together on our board.

I should mention that Jeff was generously funding many of the pieces and people within the movement we were all becoming a part of. Collectively we've been called the Medical Freedom Movement, the Health Tyranny Protest Movement, and my favorite, the Medical Dissidents. Jeff has given money to Tess Lawrie's and Bobby Kennedy's organizations, as well as hired advisors like Robert Malone and Aaron Kheriaty to consult on his own new nonprofit called the Unity Project, among others. His support and funding of the People's Convoy are legendary. I was with him at the launch of the truckers' convoy and the many days in Hagerstown, Maryland at the end. His RV was essentially the second headquarters behind Brian Brase and the rest of the leading truckers. He is also one reason many of us Dissidents survived during the thick of the war—at least financially.

I later learned that Jeff and his wife April, both deeply spiritual people, had made a commitment early on in their marriage that they would donate 50 percent of their income to charity. And they have kept that commitment. They are warriors and leaders. If anyone reading this is despairing over all that has happened and continues to happen to humanity on a societal and spiritual level, I urge you to remember there are still amazing and courageous people in the world like Jeff and April fighting for justice and sanity.

Thanks to our generous supporters, the FLCCC was finally and solidly in the black. With cash in the coffers for the first time, our potential was limitless. We could build out a respectable website, pay for legal counsel, hire support staff to do the endless list of creative and organizational things a bunch of white coats most certainly could not do, and actually throw an occasional dollar or two to the countless folks who had until this point worked tirelessly for free.

Game on.

As things were heating up around the FLCCC, Paul and I were still working clinically full-time. In fact, the whole team—which at the time included Joyce, Betsy, Frank, Fred, Keith, Joe, Jose, and Umberto—was doing this work on the side. Our protocols were getting increasing attention from the public, more media requests were coming in, and we were all going full tilt. We knew we needed a captain, someone who could completely devote themselves to navigating the FLCCC forward.

It was Joyce's idea to ask Sean Burke to be our executive director. Sean and his team had done a masterful job managing the press conference media as well as the ivermectin documentary. Sean was passionate and connected and outspoken, with a huge brain and an even bigger heart. Sean was also the founder and CEO of the Changemaker Foundation,[3] which was dedicated to supporting and strengthening individuals and nonprofits who are actively bringing about positive change in the world.

Sean was a perfect choice—but it was a big ask. Massive, in fact. To our unbridled joy and relief, Sean agreed to drop everything else he was doing and fully commit to the FLCCC and its mission. That poor bastard had no idea what he was getting into. For the next six months, our new captain would be building an airplane while flying it.

CHAPTER EIGHTEEN

My Senate Testimony on Ivermectin

Before we dig any deeper into the war trenches, I want to give a quick—yes, quick—history of the defamed drug that is the heart of this book. Most people probably only know ivermectin as the most maligned and controversial treatment in the history of medicine . . . which is a shame, as it has a remarkable story of origin and impact.

In the late 1970s, Japanese microbiologist Satoshi Omura of the Kitasato Institute in Tokyo was studying various microorganisms in the hopes of identifying promising medicinal compounds. He and his team collected thousands of soil samples from around Japan and screened them for therapeutic potential. It was on one of these excursions, while visiting a golf course near the coast of Honshu, that Omura unearthed an unusual Streptomyces bacterium.

Omura sent the sample to an American veterinary scientist and parasitology specialist, William Campbell, at Merck Research Laboratories in the US. Campbell found that the bacterial culture contained active compounds which could cure mice infected with parasitic worms; they named these active compounds avermectins. Ultimately the team developed the compound into a safer, more tolerable, and more effective drug—ivermectin—that wards off parasites in animals *and* humans.

What I find fascinating is that despite decades of searching around the world, the Japanese Streptomyces strain remains *the only source of avermectin (ivermectin's precursor) ever found*. Had Professor Omura been born anywhere else in the world or decided to sample the soil from a baseball field instead of that golf course, ivermectin would likely never have been discovered.

Ivermectin has been called a "wonder drug" for decades due to its effectiveness and safety in treating a range of infections in humans and animals, and because it has nearly rid the world of two of its most disfiguring and devastating diseases, river blindness and elephantiasis.

What's more, in 1987, Merck & Co partnered with the WHO to establish the Mectizan Donation Program, which pledged to donate ivermectin *for as long as was needed* to control river blindness and eliminate the disease. (This was clearly before Pharma had gone Harma.) This program provided more than 57 million treatments in its first twenty years alone—and prevented untold suffering and countless deaths around the world. These actions were exactly in line with the vision of George W. Merck, the son of the company's original founder. Speaking to the Medical College of Virginia at Richmond in 1950, the junior Merck declared, "We try to remember that medicine is for the patient. We try never to forget that medicine is for the people. It is not for the profits. The profits follow, and if we have remembered that, they have never failed to appear."[1] Although such sentiments must seem absurd in the modern age, apparently, they meant something back then.

Ivermectin's tremendous global impact solidified its spot on the World Health Organization's List of Essential Medicines and earned Campbell and Omura the Nobel Prize in Medicine in 2015 for its discovery. (Can you win a Nobel Prize twice for the same find? Asking for a friend.)

All of this is why, one month after I posted my ivermectin paper on that preprint server, I testified for the second time in a Homeland Security Committee Hearing for Senator Ron Johnson. This time, the topic was the efficacy of ivermectin in the treatment of Covid-19.

Recall that as a life-long disciple of the *New York Times*, I had been indoctrinated to despise Senator Johnson. That baseless hatred had turned into a profound respect within minutes of our first phone call back in May, and his continued advocacy for truth and integrity have only deepened my admiration. I wouldn't hesitate to call him a hero, as I believe that his outspokenness and efforts have saved countless lives around the world.

Senator Johnson gathered a group of Covid dissident all-stars on Zoom for a series of hearings, including the tenacious, unshakable Dr. Peter McCullough, eminent epidemiologists Dr. Harvey Risch and Dr. Jay Battacharya, and the extraordinary Dr. George Fareed, who by that time, along with Dr. Brian Tyson, had treated several thousand Covid patients with nary a hospitalization and not a single death. We discussed strategy around who would testify and for how long and on what. I would have described that meeting as electric; Ron called it, "like herding cats."

The first hearing featured a fired-up lineup including McCullough, Risch, and Fareed, who presented immense data on how our regulatory agencies and academia had gotten HCQ wrong. Although their marshaling and presentation of the supportive data for HCQ was expert and erudite, it ignited a media firestorm attacking both them and Senator Johnson while predictably celebrating the response of the willfully ignorant, odious, and servile academic from Brown University, Ashish Jha.

Jha played the classic evidence-based-maniac trafficking in pseudoscience by declaring the supportive evidence as "low quality," "insufficient" or "conflicting." Note this trick is standard, as only those in authority get to determine what sufficient evidence is. As one would expect, Jha followed these bogus claims by proclaiming the fraudulent trials of HCQ in the high-impact journals to be "high-quality," and "rigorous." It was the very same narrative that would engulf ivermectin over the next two years.

The highlight of the whole hearing was when Senator Johnson asked Jha, "Have you ever treated a Covid patient?" Jha's answer was a sheepish and reluctant, "No." It was a mic drop moment, and a definitive win for Senator Johnson and the Dissidents (which would be an epic name for a band, incidentally).

Jha was later rewarded for his disinformation efforts with a position as the White House Coronavirus Response Coordinator. He would become most famous in this role for one particularly asinine statement. "I really believe this is why God gave us two arms," Jha prattled during a White House press briefing in September 2022. "One for the flu shot and the other one for the Covid shot."

That first hearing triggered a widespread smear campaign against Senator Johnson, HCQ, and me and my fellow expert panel members. The *New York Times* ran a disgraceful opinion piece, calling Johnson one of the "snake-oil salesmen of the senate." It sounded like a headline not written by a journalist, but by a PR propagandist with a clear agenda to discredit any dissenting voices. Unbeknownst to those of us testifying, we were stepping onto a battlefield that would lead to the loss of our academic careers. I can assure you, even (make that *especially*) knowing what we know now, any one of us would do it again in a heartbeat.

The second hearing was two weeks later; I was up.[2] In hindsight, it's clear to me that I didn't fully understand how threatening the ivermectin testimonies would be to pharma and the Covid establishment, who were poised to roll out their beloved "vaccines" and had their own lineup of worthless antivirals like Paxlovid and molnupiravir (brand name Lagevrio) waiting in the wings.

Prior to my testimony, I was asked to submit a written statement[3] summarizing my position and qualifications, which was entered into the Congressional Record. In it, I was already calling out the illogical censorship, the inexplicable inaction of our "health" agencies, and the damaging medical directives that were wholly divorced from science. Looking back, I nailed it—even though I was just beginning my journey of discovery into the shocking and near-total corruption of our health system. Now I see it all clearly and simply as one of the most damaging and coordinated propaganda campaigns in history. My favorite paragraph from my written statement is this:

> Numerous studies have consistently reported large magnitudes of benefits in all the disease's phases but with the most significant public health impact in the prevention of transmission. On this compelling evidence, we recommend ivermectin's administration for both prophylaxis in all high-risk patients as well as in the early and late phases of the disease. If this were to occur nationally and globally, we predict that, like in many of the regions shown above, the pandemic will end, the economy can re-open, social interactions and activity can resume, and life can normalize. The expected impact will allow our nation to grow and focus on the multitude of other pressing problems facing our society.

As we are all sadly aware, that is not even close to what happened.

Ranking committee member Senator Gary Peters opened the hearing with a statement that accused Chairman Ron Johnson and our entire panel of expert witnesses of "playing politics with public health" and "promoting unproven therapeutics." And then that spineless shill—the only Democrat committee member who even bothered to show up at all— *walked out of the hearing.* The ignorance and disrespect was shocking. I was so enraged I couldn't even think.

Luckily, I didn't testify for another hour or so, but I was still fired up when my turn came. In my spoken testimony I could only cover a portion of what I included in my written statement, even in the extended time Senator Johnson granted me, but believe me, I let it rip. It wasn't just the childish, insulting walkout that had whipped me into an emotional frenzy. I had spent the previous hour thinking about the daily horrors of the past two years; of watching colleagues and experts and health systems make insanely stupid decisions; of witnessing relentless under-treatment and needless death; of working so many hours, not only running ICUs but building the FLCCC and researching and publishing my insights into Covid. To be accused, of all things, of being a political actor? It was more than I could take.

It was probably the lone case in my life where my supposed "justifiable anger" actually led to something good. My testimony literally went viral. I say literally because the definition of a "viral video" is one that exceeds forty thousand views in four hours, and/or exceeds a million views in total. Apparently, my taped testimony surpassed both benchmarks. So much so that I got a text from my FLCCC team telling me that Fox News wanted to interview me *while I was still in the hearing room*. That video put both ivermectin and the FLCCC on a much larger, global map.

Lucky Pierre strikes again.

Since that testimony, every one of us at the FLCCC—along with every treatment provider on the planet using HCQ or ivermectin—has had our credibility attacked. No matter how many thousands of patients we treat with near perfect results, our successes are ignored, maligned, or outright dismissed. The evidence I presented in that testimony was, in my mind, both enormous and irrefutable. It would take a few more months for me to accept the reality that no amount of incontrovertible, rock-solid science would *ever* be considered sufficient as long as it was inconvenient to Big Pharma's plans or profitability.

CHAPTER NINETEEN

An Unmovable Mountain of Evidence

As of this writing, more than twenty-six months have passed since that senate hearing. The evidence base to support ivermectin is now so massive, it's genuinely incomprehensible that both the media and our so-called health agencies continue to ignore it. It's a no longer debatable fact that treating Covid with ivermectin leads to highly statistically significant improvements in time to viral clearance, time to clinical recovery, hospitalizations, and death. And yet the medicine is not part of any official Covid protocol in the advanced health economies of the world, and if you asked people on the street what they know about ivermectin, I'm betting at least six out of ten would use the words "horse dewormer" in their reply.

As of this writing, there are ninety-five studies from 1,023 scientists including 134,554 patients from twenty-seven countries that show ivermectin's efficacy.[1] (Please read that sentence again.) These stats are from c19early.org, a website that features the work of an anonymous group of expert statisticians and researchers and which has been an indispensable tool to clinicians and researchers around the world during Covid. From the onset, they established a rigid protocol for data analysis, such that their approach is consistent across all medicine evaluations, even the Big Pharma ones. They update their analyses daily as studies on emerging therapeutics appear in peer-reviewed literature or on preprint servers.

Their scientific objectivity, consistency, and comprehensiveness in compiling clinical trial evidence for every single therapeutic studied in Covid is an unparalleled body of work, and much of it is extremely inconvenient to Big Pharma's interests.

No wonder the researchers want to be anonymous.

Based on a meta-analysis of combined data (and meta-analyses have long been considered the highest level of evidence), the researchers have to date identified forty-two medicines and interventions that lead to improved outcomes in Covid—the vast majority of which have been categorically ignored by the government and media. Guess which medication sits at the tippy top of the list of medicines which have at least four trials to support? Yup, little old ivermectin, with ninety-five controlled trials (seventy-eight of which have been published in peer-reviewed medical journals) including over 130,000 patients.[2] Further, there are numerous published studies of the effects of widespread ivermectin distribution programs from Paraguay,[3] Argentina,[4,5] Brazil,[6] Mexico,[7] the Philippines, and Peru,[8] all of which reported astonishing success in decreasing hospitalizations and deaths. In other words, ivermectin is one of the most "proven" medicines in history and also one of the most studied.

In fact, in a review paper on ivermectin in Covid by Professor Omura's team and published in the prestigious *Japanese Journal of Antibiotics* in March of 2021, first author Morimasa Yagisawa reported that in the previous month, a meta-analysis was carried out on 14,906 patients in forty-two clinical studies (including twenty-one randomized controlled trials with 2,869 patients). The data showed improvements of 83 percent in early treatment, 51 percent in late treatment, and 89 percent in preventing disease onset. Based on the results of these forty-two trials, Yagisawa concluded that the probability that the judgment of the superior clinical performance of ivermectin is wrong is estimated *at one in four trillion.*[9]

Where things get interesting is when you look at the therapies that *have* been recognized or recommended by the US government to treat Covid.

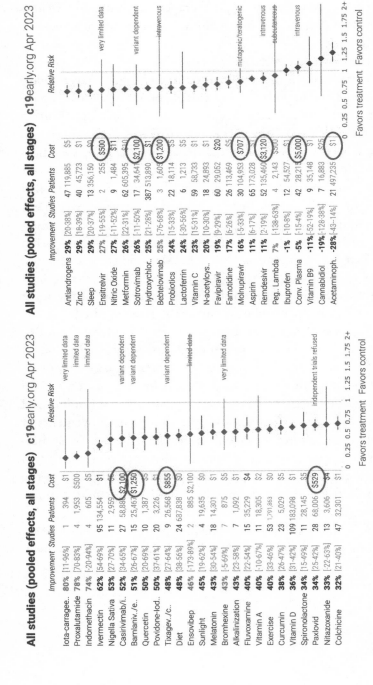

Figure 7. Forest plot showing estimated treatment benefits of forty-nine therapies tested in Covid-19. Darker shaded triangles indicate therapies with at least four clinical trials completed. The horizontal line through each triangle reflects the 95 percent confidence interval. Any line that crosses the center vertical line indicates the estimate is non-statistically significant. Large triangle in bottom shaded area reflects the pooled estimate of benefit. Circled are the costs of the only therapies recommended by the U.S government during Covid. (Source: c19early.org)

Nearly every one of them is a patented, exorbitantly costly, ridiculously prof-itable, Big Pharma pill or injectable.

The only generic, repurposed drug on the recommended list above is acetaminophen (Tylenol), which studies[10] show actually caused harm to the tune of a statistically significant 27 percent *increase* in mortality (unsur-prising because the need to treat fevers, which are a protective physiologic response to infection, is a pervasive myth). Similarly, but on the other end of the price spectrum, throughout 2020, the nation's top academic medi-cal centers went all-in on convalescent plasma, which comes with an eye-popping average price tag of $5,000 for a single treatment. The best estimate is that it was associated with a 36 percent *increase* in mortality if given early and a 5 percent increase if given later.[11]

Up until now, I've mostly been talking about treating Covid with ivermec-tin—I haven't even gotten to its enormous prevention (prophylaxis) capabili-ties, demonstrated in seventeen clinical trials, four RCTs, and one propensity matched study. Sixteen of the seventeen trials produced not only large magni-tude positive impacts, but highly statistically significant ones as well. I have never in my career seen an evidence base this strong for any preventive therapeutic.

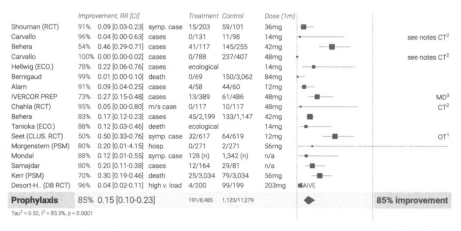

	Improvement, RR [CI]			Treatment	Control	Dose (1m)		
Shouman (RCT)	91%	0.09 [0.03-0.23]	symp. case	15/203	59/101	36mg		
Carvallo	96%	0.04 [0.00-0.63]	cases	0/131	11/98	14mg		see notes CT²
Behera	54%	0.46 [0.29-0.71]	cases	41/117	145/255	42mg		
Carvallo	100%	0.00 [0.00-0.02]	cases	0/788	237/407	48mg		see notes CT²
Hellwig (ECO.)	78%	0.22 [0.06-0.76]	cases	ecological		14mg		
Bernigaud	99%	0.01 [0.00-0.10]	death	0/69	150/3,062	84mg		
Alam	91%	0.09 [0.04-0.25]	cases	4/58	44/60	12mg		
IVERCOR PREP	73%	0.27 [0.15-0.48]	cases	13/389	61/486	48mg		MD³
Chahla (RCT)	95%	0.05 [0.00-0.80]	m/s case	0/117	10/117	48mg		CT²
Behera	83%	0.17 [0.12-0.23]	cases	45/2,199	133/1,147	42mg		
Tanioka (ECO.)	88%	0.12 [0.03-0.46]	death	ecological		14mg		
Seet (CLUS. RCT)	50%	0.50 [0.33-0.76]	symp. case	32/617	64/619	12mg		OT¹
Morgenstern (PSM)	80%	0.20 [0.01-4.15]	hosp.	0/271	2/271	56mg		
Mondal	88%	0.12 [0.01-0.55]	symp. case	128 (n)	1,342 (n)	n/a		
Samajdar	80%	0.20 [0.11-0.38]	cases	12/164	29/81	n/a		
Kerr (PSM)	70%	0.30 [0.19-0.46]	death	25/3,034	79/3,034	56mg		
Desort-H.. (DB RCT)	96%	0.04 [0.02-0.11]	high v. load	4/200	99/199	203mg	NAIVE	
Prophylaxis	85%	0.15 [0.10-0.23]		191/8,485	1,123/11,279			**85% improvement**

Tau² = 0.52, I² = 83.3%, p < 0.0001

Figure 8. Forest plot of seventeen controlled trials of ivermectin in the prevention of Covid-19 from seventeen controlled trials. Squares indicate the estimated treatment ben-efit. The horizontal line through each square reflects the confidence interval (the shorter the line reflects a more precise estimate). Any line that crosses the center vertical line indicates the estimate is non-statistically significant. Large triangle in bottom shaded area reflects the pooled estimate of benefit from all studies. (Source: c19early.org)

Compare ivermectin's meta-analysis on the previous page with molnupiravir's to see just how remarkable the above evidence base truly is:

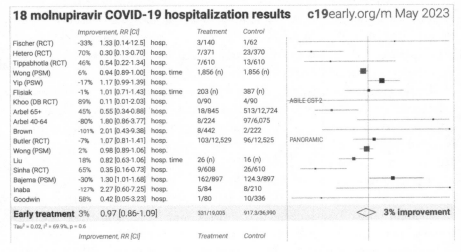

Figure 9. Forest plot of eighteen controlled trials of molnupiravir in the early treatment of Covid-19. Squares to the left of the center line indicate an estimated treatment benefit. Squares to the right of the center line indicate an estimated treatment harm. The horizontal line through each square reflects the 95 percent confidence interval. Any line that crosses the center vertical line indicates the estimate is non-statistically significant. Large triangle in bottom shaded area reflects the pooled estimate of benefit from all studies. (Source: c19early.org)

The data is so solid that the establishment (and by that, I mean the vaccine promoting machine) couldn't attack or try to discredit it without a) looking foolish, and b) drawing attention to it. So, what did they do? Simple: They ignored it.

To wit, in the WHO's updated guideline document regarding ivermectin, they include this ridiculous disclaimer: "This fourth version of the WHO living guideline addresses the use of ivermectin in patients with COVID-19. It follows the increased international attention on Ivermectin as a potential therapeutic option. *While ivermectin is also being investigated for prophylaxis, this guideline only addresses its role in the treatment of COVID-19* [emphasis mine]."

See? That was easy. Just don't look at the evidence at all. Don't publish it, don't analyze it, just pretend it's so unimportant that it is not even worth your time. After all, if you were to admit that a cheap, safe, widely available therapeutic *could in fact prevent Covid*, what do you suppose that would do to the demand for your experimental gene therapy shot?

The NIH ignored ivermectin's prophylactic promise as well. As of March of 2023, they still recommend vaccines as the "most effective" primary Covid prevention method, and even included the words "as soon as possible" in their guidelines. They find "insufficient evidence" to recommend vitamin D, and specifically recommend against monoclonal antibodies and hydroxychloroquine. What do they say about using ivermectin in prevention of Covid-19? Literally, nothing. They namelessly lump it in the category of "other drugs," about which they declared this: "Other agents have been studied for use as COVID-19 PrEP [pre-exposure prophylaxis]. The Panel recommends against their use as COVID-19 PrEP, except in a clinical trial, because the available data have not demonstrated sufficient benefit from use of the agents."[12] The old "insufficient evidence" card.

Interestingly, earlier in the pandemic, the NIH website actually contained this guidance regarding ivermectin: "Some observational cohorts and clinical trials have evaluated the use of ivermectin for the prevention and treatment of COVID-19. Data from some of these studies can be found in Table 2d."

Even when there was a Table 2d, they only included prevention RCTs, and only a couple of them at that. They simply ignored the rest of the massive amount of positive observational controlled trial (OCT) data. And then the prevention RCTs vanished from the table. They still don't even mention the study results released from a large, high quality RCT[13] done by a French biotech company that found a 72 percent reduction in infections.

Ignoring evidence is merely one tactic that's been used to discredit ivermectin. In fact, there's been a systematic and sophisticated smear campaign against what the WHO formerly deemed "an essential medicine" that reads like something out of a spy novel. At its heart is the very thing that medical freedom fighters have been (unknowingly) fighting since the beginning of the pandemic: disinformation.

CHAPTER TWENTY

A Primer on Disinformation

Having fought in the "War on Ivermectin" now for almost two and half years, I know most of the military plays. But when I first set foot on the battlefield, I was blissfully unaware of the rules of engagement. Hell, I didn't even know I was fighting in a war.

One thing *was* crystal clear to me: Something illicit was happening around ivermectin, and Big Pharma's fingerprints were all over the crime scene. But in the beginning, I truly believed that the pandemic would be over in a matter of months—just as soon as our review paper was published. The world would know that there was an incredibly effective agent to prevent and treat Covid-19; deaths would stop, and life would resume.

It physically pains me to write that last sentence.

I credit my combat training to two people, both of whom appeared in my life around the same time. The first was a man who writes under the pen name Justus Hope, MD, author of *Ivermectin for the World*. I had come across his book as well as multiple articles published in a California newspaper called *The Desert Review* in my research, so I knew who he was when he reached out. We had several in-depth conversations during which he explained his long-standing interest in Big Pharma's war on repurposed drugs. That interest was triggered by a close friend with brain cancer which led him to the discovery that there were multiple effective repurposed drugs to treat cancer that had long been suppressed by Big Pharma. Early in the pandemic, he published a book called *Surviving Cancer, Covid-19, and Disease: The Repurposed Drug Revolution*. I was beginning to understand that this was an old, old war.

My second mind-altering mentor during this period was a complete stranger named Bill Grant, PhD, a physicist and the founder and president of the Sunlight, Nutrition, and Health Research Center in San Francisco. Bill is also one of the world's foremost experts on the science behind vitamin D, with more than 300 peer-reviewed papers to his name. Out of the blue, Bill reached out to me in March of 2021 with a simple, two-line email:

> Dear Dr. Kory,
> What they are doing to ivermectin they have been doing to Vitamin D
> for decades.
> -Bill

The note was followed by a link to an article by a group of scientists detailing precisely how disinformation is used to sway public opinion. Intrigued, I clicked the link.

The article described various disinformation tactics by equating them to American football plays. By the time I got to the end of that article, a switch inside me had flipped. I instantly knew that it was the key to understanding a world that I no longer recognized.

The article went on to detail five primary disinformation "plays" or tactics used by companies or industries when science emerges that is inconvenient to their interests: the fake, the fix, the blitz, the diversion, and the screen. As I read, I could think of dozens of examples *for every single one* of those maneuvers that had occurred around ivermectin since my senate testimony had gone viral.

The mother of all Macy's 4th of July fireworks celebrations was going off in my brain; one realization exploding after another, each one brighter and more astonishing than the last.

Holy crap. The FLCCC was in the middle of a disinformation war with the pharmaceutical industry.

From that day on, that conceptual framework was the only thing that could make sense of what had happened and what was yet to happen in my attempts to highlight one of the safest and most effective treatments in any disease in history.

Although each play was widely represented in the events surrounding the Covid response, "the fake" was by far the most prominent—and the most damaging. In regard to repurposed drugs specifically, it involves conducting trials "designed to fail," selectively publishing negative results while censoring positive results, and planting negative ghost-written editorials in legitimate journals. The article emphasized that these tactics can gravely undermine public health and safety.

You don't say.

"The fake" formed the foundation of a campaign that would result in one of the most significant humanitarian catastrophes in history, causing millions of deaths around the world.

To be clear, ivermectin wasn't the first casualty of World War Covid. The same tactics had been used against hydroxychloroquine (HCQ) in 2020 and had they not, HCQ would have been deployed at the onset of the pandemic and saved even more lives. The closest and best description of *that* war I've discovered was featured in Robert F. Kennedy's *The Real Anthony Fauci* (Skyhorse Publishing, 2021), a brilliant, expertly researched, and undeniably incriminating takedown of "America's Doctor."

"HHS's early studies supported hydroxychloroquine's efficacy against coronavirus since 2005, and by March 2020, doctors from New York to Asia were using it against Covid with extraordinary effect," Kennedy wrote. By autumn, more than 200 studies supported treatment with hydroxychloroquine. "From the outset, hydroxychloroquine and other therapeutics posed an existential threat to Dr. Fauci and Bill Gates' $48 billion Covid vaccine project, and particularly to their vanity drug remdesivir, in which Gates has a large stake. Under federal law, new vaccines and medicines cannot qualify for Emergency Use Authorization (EUA) if any existing FDA-approved drug proves effective against the same malady."

In other words, if HCQ or ivermectin had been recognized as a viable treatment, the massive cash cow that was the global Covid-19 vaccine campaign would have been slaughtered on the spot.

Keep in mind that HCQ and ivermectin not only threatened the vaccine campaign, but also the massive and exploding competitive market for other pricey Big Pharma products like Veklury (commonly known by its generic name, remdesivir), Paxlovid, molnupiravir, and monoclonal antibodies. Never in history had two generic, repurposed medicines threatened a marketplace of such a colossal size.

The answer to that pesky little conundrum?

Disinformation.

Over and over, each devious play has been strategically deployed to further the interests of the establishment to the unbridled disservice of mankind.

CHAPTER TWENTY-ONE

Big Harma

To understand how disinformation plays out on the field, it's crucial to know a bit about the players. To be clear, the pharmaceutical industry had a well-documented criminal history long before Covid. Companies including Pfizer, Johnson & Johnson, AstraZeneca, Merck, Eli Lilly, GlaxoSmithKline, and dozens more have paid tens of billions of dollars in both criminal and civil fines for violations including failure to disclose safety data, bribing physicians and scientists, Medicare fraud, poor manufacturing practices, and making false and misleading safety statements. Our friend Pfizer sits at number two on the largest-ever-criminal-settlement list, shelling out $2.3 billion alone in fines for off-label promotion and paying kickbacks.[1]

What I'm saying is Big Pharma isn't exactly a trustworthy bunch.

In the seminal book *Deadly Medicines and Organized Crime,*[2] Peter Gøtzsche concluded that the pharmaceutical industry meets all the characteristics of an organized criminal enterprise. With the largest profit margins of any industry in the United States and as the third most profitable industry on earth (after software and computer peripherals), it should not be surprising that Pharma also commits more than three times as many serious or moderately serious law violations as any other industry. Similarly, Pharma has the worst record for international bribery and criminal negligence for the unsafe manufacturing of its products.

In a review paper published in *JAMA,* researchers found that of the twenty-six Pharma companies they included in their analysis, 85 percent had paid financial penalties for illegal activities totaling 33 billion dollars in the years from 2003 to 2016.

If you taxed yourself to come up with a list of one hundred illegal, immoral, or just plain shameful things drug companies might do to make a buck, I'd bet my last dime they've done every single one. And keep in mind, these are just the crimes they've been *caught doing and convicted for*. What do you suppose the odds are that this is the full criminal list?

Bribery is probably the most common offense. Consider: Your company makes the fourth most effective blood pressure lowering medication on the market. How are you going to get doctors to prescribe your d-list drug above numbers one through three? From fancy dinners and lavish trips to cash and lap dances, potential "incentives" are limitless.[3] They don't just give doctors kickbacks for prescribing their drugs; just about any warm body who can affect sales—hospital administrators, cabinet ministers, health inspectors, customs officers, tax assessors, drug registration officials, factory inspectors, pricing officials, even political parties—is fair game for a payoff.

Some of the largest fines in pharmaceutical history have been levied against companies for concealing adverse reactions and deaths caused by their drugs. Honestly, my blood pressure is skyrocketing as I type. The fact that a company can create a product that maims or kills people and their punishment is the equivalent of a parking fine is mind-blowing to me. Can you imagine a carnival ride where a handful of every few thousand riders would get tossed to their deaths . . . *and the ride could continue to operate (after they paid this nice fee)?* It's bananas.

Of course, there's more. Pharma companies overcharge low-income patients and state health agencies for medication and then conceal that activity from the government. They routinely lie to the federal government about *all* their illegal activities, and if or when they get busted, they pay another "whoops!" fine—and their executives often get promoted. Just like in the Mafia. I wish I were joking.

Peter Rost, MD, former vice president of Pfizer, wrote an epically titled book called *The Whistleblower: Confessions of a Healthcare Hitman*.[4] In it he writes,

> It is scary how many similarities there are between this industry and the mob. The mob makes obscene amounts of money, as does this industry. The side effects of organized crime are killings and deaths, and the side effects are the same in this industry. The mob bribes politicians and others, and so does the drug industry. . . . The difference is, all these people in the drug industry look upon themselves—well, I'd say 99 percent, anyway—look upon themselves as law-abiding citizens, not as citizens who would ever rob a bank. It's almost

like when you have war atrocities; people do things they don't think they're capable of. When you're in a group, people can do things they otherwise wouldn't, because the group can validate what you're doing as okay.

Mobsters are famous for paying off law enforcement officers, public officials, witnesses, and juries. It's amazing how many friends even the dirtiest of dogs can buy. Big Pharma is no different. As the largest lobbyist on Capitol Hill by miles, Pharma spends three times what the entire oil and gas industry does promoting their agendas. If you divide the total amount spent on lobbying ($356.6 million) by the number of members of Congress (535), it averages out to $665,000 per member.[5] Note these figures refer to just a single year (2021) of spending.

It's harder to find the dissimilarities between a predatory corporation and an organized criminal enterprise than it is to find the equivalence. Gøtzsche, also a founder and former board member of the Cochrane Library whose integrity got him ousted, says this of pharmaceutical industry devolution:

"The crimes are so widespread, repetitive, and varied that the inescapable conclusion is that they are committed deliberately because crime pays. The companies see the fines as a marketing expense and carry on with their illegal activities, as if nothing had happened."

Pharma may brush off the carnage as collateral damage, the cost of doing business, or just another unfortunate coincidence, but I won't. Something needs to change. The actions of these massive, affluent, liability-immune corporations are directly responsible for the human catastrophe we've been living through for the past three plus years. Their crimes are the reason I enlisted in this army, and the impetus for writing this book.

Author Christine Cain said, "It is one thing to be awakened to injustice and quite another to be willing to be inconvenienced and interrupted to do something about it." It is my life's greatest hope that what you're about to read nudges you gently into the latter category.

CHAPTER TWENTY-TWO

The First Skirmish

The very first Big Pharma attack in the war on ivermectin occurred in April of 2020, soon after a groundbreaking paper by Leon Caly and Kylie Wagstaff out of Monash University in Australia showed that *SARS-CoV2 essentially disappeared from a cell culture within 48 hours of being exposed to ivermectin.*

In a truly historic response to the gravity of the situation in Peru, many Peruvian cities incorporated ivermectin into the national protocol based solely on this in-vitro study. *TrialSiteNews,* one of the bright lights in the landscape of Covid media darkness, produced a short but powerful documentary[1] on the Peruvian program. As the documentary reveals, ivermectin did not stay on the protocol very long, as Peru's big city academics revolted, arguing that it was "reckless" to put a medicine on a national protocol without a large RCT supporting it.

Americans were getting wind of ivermectin as well, causing a brief run on the veterinary version. On April 10, 2020, the FDA issued this warning:

> FDA is concerned about the health of consumers who may self-medicate by taking ivermectin products intended for animals, thinking they can be a substitute for ivermectin intended for humans. Please help us protect public health by alerting FDA of anyone claiming to have a product to prevent or cure COVID-19 and to help safeguard human and animal health by reporting any of these products.[2]

What world were we living in? "It's your civic duty to narc on your neighbor if you suspect he's taking any horse paste." Sadly, I'm sure lots of folks thought exactly that.

Two months later, the first Pharma hit men struck. A researcher named Craig Rayner published a letter to the editor in *Antiviral Research*,[3] the same journal which had published the groundbreaking Monash study. (Rayner would show up again later as one of the investigators of the TOGETHER trial which would tap the final nail into ivermectin's coffin.) He and his team nearly tripped over themselves in their rush to bash Wagstaff's findings, writing:

"Caly et al. at Monash University in Australia recently published a paper in *Antiviral Research*, reporting that ivermectin, a medication widely used for the treatment of certain parasitic diseases in humans and livestock animals, inhibits the replication of SARS-CoV-2 in cell culture (Caly et al., 2020).[4] Despite the authors' cautious conclusion that ivermectin 'warrants further investigation for possible benefits in humans,' the paper has excited widespread interest on medical and veterinary websites, *which often incorrectly describe the drug as a treatment or cure for COVID-19.* These inappropriate statements led to a warning by the US FDA, that ivermectin in veterinary products should not be used for human therapy."

Alert the press! The FDA says veterinary medications should not be used by people!

Pardon my French, but no shit.

The Monash study was a huge global event. It was remarkable that an in vitro study got that much attention and led to Peru incorporating ivermectin into its treatment guidelines. Which is precisely why the system shills needed to attack swiftly and sent in Rayner and his coauthors from his Pharma employed company to do damage control. The database c19early.org had this to say about Rayner's conflicts in the later TOGETHER trial:

> One of the senior investigators was Dr. Craig Rayner, President of Integrated Drug Development at Certara —another company with a similar mission to MMS Holdings (helping pharmaceutical companies get approval and designing scientific studies that help them get approval). They state on their website that: "Since 2014, our customers have *received over 90% of new drug and biologic approvals by the FDA*." One of their clients is Pfizer.[5]

Although I singled out Rayner because of his involvement with the later TOGETHER trial, there were a total of six authors of that letter, five of whom worked for Certara.

Rayner's letter planted the seed for the first anti-ivermectin "narrative," which was that *standard dosing could never achieve the effective concentrations reached in the Monash University study.* This was total nonsense. But it was the first propaganda campaign that we were forced to fight against.

Around this time, Professor Satoshi Omura, the Nobel Prize winning codiscoverer of ivermectin, politely wrote to Merck (the manufacturer) for funds to study its clinical efficacy in Covid. Recall that there was nearly a decade of studies showing ivermectin's ability to halt the replication of over ten RNA viruses. In his Nobel Prize acceptance speech in 2015, Omura called ivermectin "the wonder drug," not only due to its incredible safety as an anti-parasitic in humans, but also its broad anti-viral and anti-tumor properties.

Merck, for reasons that sound a lot like *off-patent-little-profit*, replied that they were not interested in supporting research into ivermectin for Covid.

In a paper Omura later published on the efficacy of ivermectin in Covid, he wrote,

> Kitasato University, based on the judgment that it is necessary to examine the clinical effect of ivermectin to prevent the spread of uncertain COVID-19, asked Merck & Co., Inc. to conduct clinical trials of ivermectin for COVID-19 in Japan. This company has priority to submit an application for an expansion of ivermectin's indications, since the original approval for the manufacture and sale of ivermectin was conferred to it. *However, the company said that it had no intention of conducting clinical trials* [emphasis mine].

So, Pfizer (Rayner et al.) tried to neuter the Monash study, while Merck refused to study ivermectin from the get-go. This was the first sign of what was to come.

Remember, Pharma creates customers, not cures.

Despite Rayner's efforts to cancel the Monash study, further evidence supporting ivermectin began to emerge. Soon after Rajter's paper was published in *Chest* in October of 2020, even more compelling clinical evidence came out of the Dominican Republic. I later got to meet the senior author of the study, Dr. Jose Redondo, a prominent cardiologist who owns hospitals, clinics, and urgent care centers down there. Over drinks, he told me "the rest of the story," as Paul Harvey used to say.

One night in March of 2020, Jose got a call from his hospital. One of his doctors had just admitted an overweight, diabetic tourist with progressive hypoxia (insufficient oxygen at the tissue level). The patient was deteriorating rapidly, and the doctor told Jose that he had heard that ivermectin might be effective and wanted to try it. Jose held a group conference call with the doctors on the treatment guidelines committee, and they agreed it was reasonable to try it. When Jose called the doctor back to tell him the news, the doctor replied, "Thanks, I already gave it to him an hour ago."

Within twelve hours of receiving ivermectin, the patient demonstrated a rapid and robust improvement in his oxygenation. Dr. Redondo and colleagues (even if they didn't know the mechanisms at the time), rapidly established a protocol, and by November of 2020, an early draft of their paper was posted on a preprint server.

In all, 2,706 patients presenting to the emergency room with Covid symptoms were given ivermectin and sent home. Just sixteen of them returned needing hospitalization, and only two of those patients died. An incredible 99.3 percent of all the symptomatic patients arriving to the ER, who were inevitably far sicker than stay-at-home patients, avoided hospitalization and death in that first crushing Wuhan wave.

One month later, in December of 2020, a paper was published in a French dermatology journal[6] about a nursing home that had suffered an outbreak of scabies (which is why it was in a dermatology journal). When scabies broke out, just before the first surge of Covid cases hit that part of France, the nursing home residents and staff were all given ivermectin. The doctors were astonished to find that in that nursing home, only 1.4 percent got Covid compared to 22.6 percent in the other nursing homes in the region. Most importantly, not a single patient in the ivermectin-treated facility needed oxygen, was hospitalized, or died. In the other homes, 4.9 percent of Covid patients died.

I'm sure it was just a coincidence.

CHAPTER TWENTY-THREE

Andy Hill and the WHO

As a result of my senate testimony going viral, coupled with increasing attention around our ivermectin review paper,[1] I was invited a week after I testified to give the opening lecture at an international conference organized by the CEO of MedinCell, a French biotech company. MedinCell routinely develops long-acting formulations of common medicines, allowing dosing to be as infrequent as every few months. Due to ivermectin's protective efficacy against malaria, they were working on a long-acting formulation that could be used to prevent the mosquito-borne infectious disease. They were also interested in the possibility that a durable, injectable form of ivermectin could act as a sort of "vaccine" against Covid.

The conference featured a dozen lecturers from all over the world. Among them were Dr. Kylie Wagstaff of the globally groundbreaking SARS-CoV2 cell culture study of ivermectin out of Monash University in Australia, several principal investigators of ongoing Covid trials, and Dr. Andrew "Andy" Hill, a researcher jointly sponsored by the global health agencies Unitaid and the World Health Organization (WHO).

Andy lectured on the third day of the conference. I missed his talk (I was battling to get our paper ready to submit for publication), but Paul was watching remotely. When he told me that a guy from the WHO had presented a systematic review of all the RCT data on ivermectin in Covid, I was shocked. I thought our group was way ahead of everyone in our compilation of that data.

I wrote to the MedinCell CEO and asked for Dr. Hill's slides and contact info. After receiving the slide deck, I was blown away as it contained markedly positive RCT results (and *only* RCT results, as that single study design was the sole scope of Andy's work), some of which I was not even aware of. His meta-analysis at the time included eleven studies which were mind-blowingly positive in terms of reduced time to viral clearance, time to clinical recovery, need for hospitalization, and death. At the time, his trials included more than 1,400 patients. It's worth noting that both Paxlovid and remdesivir were awarded Emergency Use Authorizations based on far fewer patients from just one RCT each. I probably don't need to point out that one positive study is far less impressive or persuasive than a dozen. The only reason you'd stop at one is if that single study produced a result that Pharma was pretty sure they couldn't replicate with further examination.

Andy responded immediately. We had an incredibly positive conversation, as any two researchers would when they've stumbled upon data that potentially has global, historic impact. We started sharing our stories of how we had "discovered" the phenomenal efficacy of ivermectin in Covid.

Andy's story went like this: He had been hired in June 2020 by Unitaid, an international health care organization funded largely by the Bill & Melinda Gates Foundation (BMGF) along with several other countries; BMGF is also the second largest funder of the WHO after the United States. Unitaid was collaborating with the WHO on the ACT Accelerator program, which was completely run and staffed by BMGF.[2] Andy was in charge of a research team tasked with analyzing trials of repurposed drugs for use in Covid.

This was fantastic news and precisely what I'd been shouting about in my senate testimony. Our governmental health agencies were not initiating a coordinated effort to identify effective, readily available drugs they could repurpose to fight Covid. And here was Andy, the head of a team doing just that at the global level!

Andy explained he'd been researching all manner of repurposed medications since June 2020 but so far none had shown efficacy in Covid. Then in November 2020, right around the time my paper hit the preprint server, Andy told me that a *professor at his university* (this will be important later) suggested he look into ivermectin. He mentioned that his project scope specifically excluded looking at the RCTs of ivermectin used in prevention, which I found odd. I mean, why would they tell him to not look at those prevention trials? There were three completed RCTs at the time, all astoundingly positive. They must have had their reasons, I recall thinking, although for the life of me I couldn't presume what those reasons might be.

Soon after my senate testimony, a former Texas Health Commissioner named Reyn Archer had reached out to me. Reyn was working as chief of staff for Nebraska Congressman Jeff Fortenberry, who was on the committee that oversaw the budgets for the health agencies. Reyn and Jeff convinced the NIH treatment guidelines committee to allow me and Paul to present our findings. The meeting was coming up in January, and we decided to invite Andy to present his data with ours. In addition to his expansive RCT data, we felt having someone on our team who worked for an international health-care agency would increase our credibility.

Little did we know that meeting would be the first battle between the FLCCC and Anthony Fauci in what has become a bloody two-plus year war.

On January 6, 2021, Paul, Andy, and I teamed up to give a twenty-minute presentation to the NIH guidelines committee. Andy essentially gave the same presentation he'd given at the MedinCell conference three weeks earlier. I presented the epidemiological analysis paper by Juan Chamie, Jennifer Hibberd, and David Scheim which showed massive reductions in both cases and deaths in the wake of Peru's magnificent ivermectin distribution program. Paul presented newly released and as-yet-unpublished experimental data from Caly and Wagstaff of Monash University which found that standard doses of ivermectin do in fact reach effective anti-viral concentrations in lung and fat tissue. Paul's data was critical, because it effectively refuted a false media narrative to the contrary that had surfaced in the wake of my testimony.

The academics and health bureaucrats on the committee were full of questions, skepticism, and dismissiveness, especially in regard to the results of the Peru program that I had presented. I suppose this was to be expected; essentially, it's their job. But not one showed even mild enthusiasm or optimism either.

At the very end of the meeting, after the discussion and question session, Alice Pau, a pharmacist in charge of coordinating the meeting, asked if we had any questions for the committee. Paul made a bold plea for the NIH to recommend ivermectin based on its unparalleled safety profile and the data we had presented.

It was my turn to ask a question. Our game plan going in was to be deferential and collegial and in no way confrontational, but I couldn't help myself. I called out one of the many elephants parading about the room.

"Of all the medicines currently being used or studied for the treatment of Covid, all have had either weak or neutral recommendations for use in the form of insufficient evidence to recommend or not recommend. Yet since August of 2020, ivermectin is the *only medicine which has been given a negative recommendation* to not use outside of clinical trials. Can I ask why that is?"

The seconds ticked by. Finally, Alice began answering a question I had not asked, so I cut her off and repeated my question. Another long pause ensued. There were over twenty "experts" on that Zoom call, and not one could cough up an explanation. The silence got so long and uncomfortable that I finally said, "Okay, I guess no one knows why."

It was rude but I was pissed.

"Come on, guys," Chairman Cliff Lane finally said to the group. "We have to answer the question." When no one else was forthcoming, Cliff pulled an Alice and began to give a similar, condescending answer to yet another question I had not asked, launching into an explanation of the grades and strengths of recommendations, of which I was expertly familiar.

They simply had no answer. The call ended awkwardly and at least on my part, angrily.

Later I would talk about the deafening silence that followed my plea with my friend Del Bigtree. Del is a TV and film producer and founder of the Informed Consent Action Network (ICAN), an advocacy group devoted to investigating the safety of medical procedures and products including vaccines, as well as educating people about their right to informed consent. Del is a public health hero who has been battling medical tyranny, funding medical studies, and creating powerful media campaigns to inform and educate the masses. In other words, he's a total badass. When I told him about what happened on that Zoom call, Del shared the story of a meeting he and Bobby Kennedy Jr. once had with an NIH vaccine committee populated by all the NIH vaccine program heads. Del and Bobby asked if any of the childhood vaccines had a placebo-controlled trial to support its use. One member replied that they had done Phase 1 and 2 placebo trials. (Clinical trials go through four phases; phases 1 and 2 only include testing in a small number of humans. Phases 3 and 4 are where large-scale trials of safety and efficacy occur.) When Del and Bobby asked if they could kindly see the trials data, no one responded. Del and Bobby sat there, waiting. After a long, uncomfortable silence—which we've since dubbed "the NIH pause"—Dr. Fauci spoke up.

"We don't do placebo-controlled trials with childhood vaccines because they are unethical," Fauci explained. I am not making this up.

Several weeks after we presented to the NIH committee, out of the blue, their guidelines changed. They no longer recommended ivermectin "against use outside a clinical trial"; instead, they quietly changed it to a neutral recommendation. You know, the one where "there is insufficient evidence to recommend or not recommend" this therapy.

It was a tiny nod to our presentation, but hardly a big win. The different strengths of recommendation—weak, moderate, or strong—can be based solely on observational controlled trial (OCT) data. OCTs look "backward" at the records of patients who took a certain medication and those who did not and then compare the outcomes of each group. The unparalleled safety profile of ivermectin combined with the consistently positive and statistically significant RCT data (let alone the dozens of OCTs) should have led to at a *minimum* a weak recommendation. But if that had happened, doctors around the world would have started treating all of their Covid patients with ivermectin—a simple shift that would have saved countless lives and effectively ended the multibillion-dollar vaccine campaign. Pharma-boy Fauci was not about to let that happen.

Around this time Paul and I connected with another well-known and respected researcher and physician named Tess Lawrie. Tess is a South African like Paul living in the UK, and she is brilliant and courageous. She has worked for decades as an expert reviewer of medical evidence and has contributed to the development of treatment guidelines for the WHO, the UK's National Health Service, and other national and international health agencies. She has published reviews in the top medical journals and databases, including many for the Cochrane Library, once considered the gold standard of such analyses—an honor subsequently lost in the wake of a very public scandal and mass resignation within its governing board (curiously timed, in my opinion, after BMGF began donating to Cochrane).

Intrigued by my video testimony, Tess began to dig into the published and posted trials data on ivermectin in Covid. Experts generally rely on what is called "pattern recognition" in their analyses; in Tess's review of the evidence, she recognized a remarkable pattern of results in the form of consistent, often large-magnitude benefits in time to recovery, hospitalizations, and death from countries and centers around the world.

So now, in addition to five highly published clinicians and researchers from within the FLCCC, there was a second independent, profoundly expert researcher acknowledging the same data signals. Tess was so impressed with the strength of the data that she recorded a video[3] pleading with then Prime Minister Boris Johnson and the UK parliament to review the overwhelmingly positive science around ivermectin. She uploaded the video to YouTube where it was almost immediately taken down, one of the first distressing acts of censorship in the campaign to suppress the evidence of efficacy of ivermectin.

After we saw Tess's video, Paul and I scheduled a Zoom meeting with her. We were wildly impressed with her thoughtfulness, dedication, and decades of experience and connected her with Andy so they could share data and insight. Tess invited Andy to collaborate on a systematic review and meta-analysis for the Cochrane Library, a a medical database of independent research that had published many of her expert reviews in the past. There were masses of people dying during the winter surge and time was of the essence. Tess knew that a review supporting ivermectin published in the Cochrane Library would get the immediate attention of doctors around the world.

Andy agreed to collaborate on the review. He would be extremely important to such an effort because the scope of his work for Unitaid was to search the clinical trial registries from all over the world to find any registered RCT on ivermectin in Covid. He had already found fifty-nine registered RCTs and was having bi-weekly meetings with the principal investigators. He was getting trial results long before any manuscript was being posted or published and sharing the data with me and Paul. It was incredible hearing about positive results before anyone else in the world.

On January 16, 2021, just ten days after our NIH guideline committee presentation and without alerting us in advance, Andy posted his meta-analysis of the ivermectin RCTs on a preprint server. Preprint servers were critical during Covid because researchers could post their data and have it publicly available without waiting months for peer review and publication.

The data presented in Andy's paper was beyond compelling. He reported that ivermectin was associated with reduced inflammatory markers, faster vital clearance, significantly shorter duration of hospitalization and an incredible 75 percent reduction in mortality in moderate or severe infection.

But there was a problem.

A devastating one.

Andy's conclusions did not match these data.

When I finished reading, the earth seemed to stop spinning. Time stood still. Nothing made sense.

For the first time in my career, I found myself reading a scientific manuscript by a researcher presenting profound and compelling data and then reaching a conclusion that actually *argued against the findings*. If scientists and researchers tend to do anything when publishing original work, it's to *over-interpret* the potential importance and impact of their data. But Andy had done the exact opposite. It simply made no sense.

In addition, his analysis was poorly written, with repeated expressions of the limitations of the data including false statements about how effective concentrations could not be reached with standard dosing (a claim we knew Andy knew was false).

Something was hugely, horribly wrong.

Paul and I swiftly wrote to Andy with our concerns and provided him with a complete peer review of his paper containing our many comments and recommendations for changes. We demanded that he take down his paper and implement the suggested revisions to be consistent with the existing data. Among other requests, we asked that he remove the statements about how effective concentrations could not be reached in the blood with standard doses. Further we called out the numerous irregularities in his paper including the repetitive citation of the "limitations" of the data presented. (The standard format for limitations is a simple list at the end of a paper; this paper had so-called limitations sprinkled throughout.)

Andy knew the importance of his data. In several public lectures he would later give that month, in contrast to the wording of the preprint paper, he was as enthusiastic about and supportive of ivermectin as Paul, Tess, and I were. In an interview with the French publication *Bon Sens*, he even emphasized the dose-dependent effects of ivermectin. Note that a "dose dependent relationship" is one of the strongest pillars of scientific support for the efficacy of a therapeutic, defined as measuring greater benefits as the dose is increased. This is what Andy said publicly at the time:

> We are seeing very clear antiviral effects. We see smaller effects when the drug is given for one day, then in dose-ranging studies, we see more and more of an effect. And then if the drug is given at a high dose, for five days, we see the strongest effect. So, how could that be happening if the drug does not stop the virus from replicating? It simply does. It does. And we've got the evidence to prove it . . . It's just a matter of time before it gets approved.[4]

Tess knew something shady was going on, and like me and Paul, believed that someone had altered Andy's conclusions and his analysis. She requested a Zoom meeting to discuss, and Andy agreed.

Tess recorded the call, and it is highly disturbing. Andy flat-out admitted that his "sponsors" Unitaid and the WHO had influenced the conclusion of his paper. Andy would not name names, but it goes without saying that whoever had altered that paper was not listed as an author. This was clear scientific misconduct. Tess was reluctant to release the recording initially but later decided it was too important not to reveal. She included the most relevant parts in a powerful documentary called, "A Letter to Andrew

Hill,"[5] which can be found online at oraclefilms.com and other uncensored streaming platforms.

After that Zoom meeting, Tess never spoke with Andy again—but I did. I foolishly thought I could get Andy to revise his preprint by playing "good cop" to Tess's "bad cop." Plus, he was still feeding me data reports from the trialists he was communicating with—and he promised me that when his contract with Unitaid ended a few months later on April 1, 2021, he would make everything right by publishing an updated meta-analysis not only independently, but in a top journal (he was highly published and knew editors of interested journals).

Meanwhile, Tess quickly formed an independent expert panel, inviting dozens of medical researchers and practitioners from around the world. She called it the British Ivermectin Recommendation Development group, or BIRD International. After meeting to review the evidence, this esteemed group of specialists nearly unanimously recommended that ivermectin be used in the prevention and treatment of Covid worldwide.[6] Tess then sent the final, comprehensive, detailed report of their findings and recommendation to every major public health agency in the world.

BIRD International initiated a massive people-powered campaign with tens of thousands of followers and dozens of affiliate organizations. Their efforts led to the creation of brand-new health-focused entities such as The International Ivermectin for Covid-19 Conference, World Ivermectin Day and not least, the broad international reach of Tess's new organization called the World Council for Health. Although BIRD International helped bring knowledge of ivermectin's efficacy to the world, the captured global media ignored BIRD as they had ignored FLCCC all along.

(Much later, in March 2023, Emmy Award–winning investigative journalist Sharyl Attkisson would report that when Janet Woodcock, the former interim head of the FDA, received the BIRD report, a flurry of emails occurred between her, Anthony Fauci, and Francis Collins, then director of the NIH. Obtained through a Freedom of Information Act (FOIA) request, each and every email is fully redacted. Shocker.)

At the time Andy's preprint bomb dropped, Paul and I were in frequent communication with a man named Lynden Alexander. Lynden was investigating all the shenanigans around the lack of a recommendation early in the pandemic for the use of corticosteroids in severe, hospitalized Covid patients—exactly what I was calling for in my 2020 senate testimony. Lynden is a wickedly intelligent forensic science communication expert and one of his many skills is performing linguistic analyses. He began to scrutinize the language, grammar, and writing style of the preprint version of Andy's paper, without knowing Tess had a different version from days earlier. Even without that knowledge, he was convinced there were multiple "voices" appearing within the paper. For at least one of those voices, English was a second language. Further, based on the language used, Lynden knew that this person was involved in regulatory activities. Lynden diagnosed this in January of 2021. It would be more than a year before we learned who the mystery author was.

It was also later revealed that Andy had been discussing his paper with Dominique Costagliola, a French epidemiologist and biostatistician. In an interview with journalist Phil Harper, Andy admitted that not only had he been communicating with Dominique, but that she had been advising him in some way. Twitter users quizzed her about this, and she confirmed it there, too.

Phil would become a central character in the ivermectin saga from this point forward. Phil is a polymath with a diverse background of interests and accomplishments having worked in journalism and documentary filmmaking, among other pursuits. He was a UK citizen and had been living in India during the early pandemic where ivermectin was in widespread use. Phil was shocked when he returned to the UK in 2021 and found a country that not only had no early treatment strategy but was also attacking, suppressing, and legislating against ivermectin. His Substack is called "The Digger" and it is masterful; I cannot recommend it highly enough. So what did Phil find out about Dominique Costagliola?

1. She was the deputy director of the Pierre Louis Institute of Epidemiology and Public Health in France.
2. She spoke English as a second language (which is what Lynden Alexander's forensic analysis had identified as a characteristic of one of the "voices" that appeared in Andy's paper).
3. She had an early history of attacking ivermectin within ten days of my testimony, as evidenced in an article titled "Coronavirus: No, ivermectin is not a 'very effective' drug against the disease."[7] That article essentially laid out the narrative refrain we hear to this day: that the trials were are small and low quality, and that "proper, large and rigorous" (i.e., Pharma-controlled) trials were needed to validate the findings.
4. She could not be considered unbiased, as she routinely received money (in the form of lecture fees, personal fees, and travel and meeting expenses) from nearly every corporation with a competing product against ivermectin including Janssen, Gilead, Merck Sharp & Dohme, Viiv, Innavirvax and Merck Switzerland.

In March of 2021, Dominique tweeted something odd. "Despite the encouraging trend this existing database demonstrates," she wrote, "it is not yet a sufficiently robust evidence base to justify the use or regulatory approval of ivermectin."[8] It's worth noting that the wording in her tweet is nearly identical to a sentence that was inserted into the conclusion of Andy Hill's preprint paper. It's also worth noting that "regulatory approval" is not something any clinician or researcher thinks about when publishing research. And Dominique's subsequent call for "large, well conducted trials" is exactly how Pharma would put the genie back in the bottle.

What Phil discovered next is what I consider the "Scoop of the Century." (It's a fitting title, given that I call what these people did to ivermectin the "Crime of the Century.") Phil analyzed the metadata embedded in the PDF file of Andy's preprint paper and discovered that it was finalized on the computer of Professor Andrew Owen of the University of Liverpool in the days leading up to the posting. It had to have been the same professor who had suggested Andy "look into ivermectin" in November of 2020. Was Big Pharma monitoring preprint servers during Covid? If I was a betting man, I'd take that straight to Vegas.

Not only was Andy's paper doctored on Owen's computer, but the professor had insane conflicts of interest against ivermectin. In fact, Dominique Costagliola's pale in comparison. To wit:

1. Owen studied molnupiravir for Merck, a direct competitor to ivermectin.
2. Owen received research funding from ViiV Healthcare, Merck, Janssen, Boehringer Ingelheim, GlaxoSmithKline, Abbott Laboratories, Pfizer, AstraZeneca, Tibotec, Roche Pharmaceuticals and Bristol-Myers Squibb.
3. Owen received consultancy fees from Gilead (another drug whose market ivermectin would have decimated).
4. Owen was the Project Lead of the Center for Excellence in Long-Acting Therapeutics Program (CELT) at the University of Liverpool. CELT received a staggering $40 million from Unitaid on January 12, 2021. (The importance of this financial influence is incontrovertible. Just let it sink in for a moment). CELT studies ways to use lipid nanoparticles in pharmaceuticals, which is a foundational technology of the Covid mRNA vaccines. Further, the Unitaid grant was shared with a spinoff start-up company in which Andrew Owen was the top shareholder.

Phil Harper found it possible that the grant agreement may have given Unitaid a say in the conclusions of any research it commissions. Phil's FOIA request for a copy of the grant agreement was denied, so I suppose we will never know. But we know.

The importance of what I am about to reveal is inestimable in its impact on depriving the *entire world* of access to and use of a life-saving drug to treat Covid: Andrew Owen was also given the responsibility for preparing the evidence base upon which the World Health Organization would make their recommendation to not use ivermectin "outside of a clinical trial" on March 31, 2021.

A professor swimming in financial conflicts of interests with pharmaceutical companies that had products directly competing with ivermectin in the now global "Covid marketplace" led a team of experts taxed with the job of assessing the ivermectin evidence for the most powerful health-care organization in the world. As if that is not absurd enough, Owen's conflicts of interest are not mentioned in the WHO recommendation document. Instead, what appears is this:

> Web searches did not identify any additional interests that could be perceived to affect an individual's objectivity and independence during the development of the recommendations.

What an absolute travesty. Individual researchers are obligated to report conflicts of interest voluntarily; it's basic, baseline scientific integrity. I have never *ever* heard of looking for conflicts of interest using "web searches." I have effectively run out of words to describe the nefariousness of these actions. Clown World had erected a spiffy new big top.

What Owen and his colleagues at the WHO did to suppress the evidence of ivermectin's efficacy in their Therapeutic and Covid-19 Living Guideline document was so openly corrupt, I quickly started writing a white paper[9] (basically an in-depth research article) about their brazen manipulations of the evidence. I spent weeks on it, extensively detailing how they had whittled down the evidence base to as few trials as possible, using arbitrary exclusion criteria. Then they graded the handful of remaining trials that showed large, positive effects as "low quality" and a large Pharma-conflicted trial that showed no benefit as "high quality." What is fascinating is that despite the fact that even amongst the paucity of remaining trials left, a massive reduction in mortality was found. They found a life-saving medicine in Covid. It was irrefutable. Yet they stated that this conclusion was of such "low certainty" it should not be acted on. "Trust the science," folks.

It was disgusting.

The FLCCC sent out my white paper via press release, trying to disseminate it as far and wide as we could. We were the Bad News Bears trying to take down the Yankees. As always, not one major media science journalist in the world picked up the story. Not one country's health-care leadership or government dared to object.

In that paper, I showed that even in the WHO's corrupt guideline, they reported seventy deaths per 1000 in the standard-of-care treated patients versus fourteen deaths per 1000 in ivermectin-treated patients. *That was an 81 percent reduction in mortality, even higher than what Tess had found.* Let me repeat that: an 81 percent reduction in mortality. Remdesivir doesn't do that. Paxlovid doesn't do that. Molnupiravir doesn't do that. Monoclonal antibodies don't do that. And Owen had conflicts with three of these "competitors."

The chart below is from the WHO guideline document, and it makes my blood boil. Look at the "Certainty of the Evidence" column and note they graded the evidence as having "very serious imprecision." Downgrading the quality of the evidence to this degree based on imprecision is flat-out wrong when the treatment effect is so large, the outcome prevented is *death*, and the medicine is one of the safest, least expensive, and most widely available in the world. Had I had a seat at the committee table, I would have raged at this. But they don't invite people like me onto captured regulatory agency committees, because I don't take Pharma money. Never have. Never will.

Outcome Timeframe	Study results and measurements	Absolute effect estimates		Certainty of the evidence (Quality of evidence)	Plain text summary
		Standard care	Ivermectin		
Mortality	Odds ratio 0.19 (CI 95% 0.09 - 0.36) Based on data from 1,419 patients in 7 studies. [1] (Randomized controlled)	**70** per 1000	**14** per 1000	Very Low Due to serious risk of bias and very serious imprecision [2]	The effect of ivermectin on mortality is uncertain.
		Difference: **56 fewer** per 1000 (CI 95% 63 fewer - 44 fewer)			

Figure 10. Pooled mortality data from trials included in the WHO meta-analysis of ivermectin in the treatment of Covid-19.

The most absurd statement in the WHO recommendation was this one:

> Applying the agreed values and preferences, the GDG [Guideline Development
> Group] inferred that *almost all well-informed patients would want to receive
> ivermectin only in the context of a randomized clinical trial [emphasis mine]*,
> given that the evidence left a very high degree of uncertainty in effect on
> mortality, need for mechanical ventilation, need for hospitalization and other
> critical outcomes of interest and there was a possibility of harms, such as
> treatment-associated SAEs [serious adverse events].

If you skimmed over that paragraph, I beg you to go back and reread it, because this is actually how they tried to justify recommending *against* ivermectin. Essentially what they are saying is that it is the WHO Guideline Committee's opinion, based on some Pharma-conflicted academic blowhard's belief that the evidence showing an 81 percent reduced risk of dying was of "low certainty," that *critically ill patients and their loved ones would feel more strongly about helping Big Pharma perform controlled testing than not dying.* How many times can I use the words ridiculous, asinine, and absurd?

The corrupt anti-recommendation that followed read like this:

> We recommend not to use ivermectin in patients with COVID-19 except in
> the context of a clinical trial. This recommendation applies to patients with
> any disease severity and any duration of symptoms. A recommendation to
> only use a drug in the setting of a clinical trials [sic] is appropriate when there
> is very low certainty evidence and future research has a large potential for
> reducing uncertainty about the effects of the intervention and for doing so at
> reasonable cost.[10]

Note how they suggest ivermectin may actually be harmful, which is even more ludicrous—and they knew it.

Had the WHO endorsed ivermectin, even using one of the conditional or "weak" recommendations available to them, it would have changed history and saved millions of lives across the world. Pharma would not have been able to conduct their later, fraudulent RCTs (because administering a placebo would have been unethical at that point). Physicians would have adopted it globally at a time when there was no "official" early treatment option for Covid outside of maybe a dozen low- and middle-income countries that had adopted ivermectin into their national or regional guidelines. A weak recommendation would have changed history by mitigating the disastrous scale and trajectory of the pandemic. And it would have prevented us from being subjected to the subsequent humanitarian catastrophe unleashed by lethal vaccines.

If anyone reading this has even a shred of doubt that ivermectin is effective in the prevention and treatment of Covid, ask yourself why such covert and duplicitous effort was put into distorting the drug's evidence base of efficacy. If it didn't work, none of these actions would have been necessary. None. Because if it didn't work, *doctors like me would not have used it nor would we still be using it.* Every one of us would have known within our first dozen or so patients if it was doing nothing and moved on. It's not that hard to tell if a drug works in a viral syndrome whose course, trajectory, and expected outcome are well known to an experienced clinician. The late, brave Dr. "Zev" Zelenko figured out HCQ quickly and went public with his findings. Peter McCullough, Harvey Risch, America's Frontline Doctors, Thomas Barody of Australia, Dider Raoult of France, Andrea Stromezzi of Italy, and many others had similarly identified HCQ as an effective treatment option early in 2020. It took calculated, coordinated effort to destroy their work. And their livelihoods.

The analysis and arguments the WHO employed were so obviously corrupt, yet the resistance was nonexistent (with the exception of my little FLCCC white paper). And that is how we lost the biggest, most important battle in the war on ivermectin.

CHAPTER TWENTY-FOUR

A Fraud Narrative Emerges

Recall that although Andy let his Unitaid/BMGF sponsors doctor his initial preprint with language minimizing the importance of his findings, he then gave public lectures in a much different tone, speaking far more positively than he had in his paper. In his final public lectures in January of 2021, Andy was extremely enthusiastic about ivermectin's efficacy, so much so that he made three very powerful statements:[1]

> "The purpose of this report is to forewarn people that this is coming: get prepared, get supplies, get ready to approve it."

> "The probability that the measured impacts on survival is [sic] due to chance is 1 in 5,000."

> "Millions of vaccine doses were manufactured and purchased at risk by countries before efficacy was confirmed. Can we start to upscale ivermectin as well?"

But Andy's paymasters clearly didn't like what he'd had to say in South Africa. A week after that lecture, I fielded an interview request for him from the *New York Times*. Andy declined, explaining that his Unitaid sponsors had informed him that he could no longer give public statements or interviews. Curious, no?

Two months later, on March 31, 2021, the manipulated WHO guideline against the use of ivermectin came out, devastating all of us. That was also the same day that Andy's research contract with Unitaid ended. True to his earlier promise to me, he continued to work independently on his meta-analysis. No longer on the Unitaid payroll, he had no sponsors overseeing and manipulating his work. I had connected Andy with the Rainwater Foundation, a philanthropic organization that has done phenomenal work in supporting research in Covid, and they had given him a generous grant to continue this work. In early July 2021, Andy published a meta-analysis[2] that was astoundingly positive (albeit ignored across major media and our health agencies). It's fascinating what happens when you conduct science in the absence of conflicts of interest.

This analysis looked at data from twenty-three ivermectin RCTs that included 2,928 patients. All of the outcomes were statistically significant except one (lower risk of hospitalization). The data showed reduced inflammatory markers, time to viral clearance and time to clinical recovery; shorter duration of hospitalization; and most importantly, reduced mortality.

For those who don't routinely read medical studies, I cannot emphasize how striking these findings were. An extremely large evidence base, consisting of only those hallowed RCTs, showing massive reductions in every important outcome, including death. *As in, we had found a freaking life-saving drug during a pandemic where millions of lives had been lost.* If we were on any other planet or living in any other time in history, this paper would have been blasted across every newspaper and television station in the world. Instead, crickets.

Predictably at this point, mainstream (or what I now call "corporate-controlled") media across the world collectively ignored Andy Hill's newer historic, landmark paper. Consider the implications of this in context: The world's leading researcher on ivermectin, having worked for global agencies and now funded by a benevolent philanthropist, compiled almost two dozen (two dozen!) RCTs which showed overwhelming benefits on every important outcome in Covid when treated with ivermectin and then published this analysis in one of the most highly regarded infectious disease journals in the world.

If you're a sociopathic Big Pharma CEO or the leader of a drug company team whose sole objective is to seek out and destroy any science that might be detrimental to your company's bottom line, what do you do about this highly inconvenient paper? Put your immense pressure on the journal editors to retract? Destroy the credibility of the researcher, perhaps by finding or planting evidence he or she is a child molester or a drug addict? (It's been done—and exposed.)[3] That might work in the short term, but the RCTs were already out there so the data could still be compiled and published by others. It's quite a conundrum.

Your only real option is to go after the evidence base itself and destroy its credibility. And how do you do that? Easy: Paint the trials included in the analyses as fraudulent. Not all or even most of them; a handful will do. Then all you have to do is craft an overarching narrative that all ivermectin science is not to be trusted. It's a disinformation strategy known as "the diversion," characterized by manufacturing uncertainty about science where little or none exists. It only takes a small seed of doubt to neuter the truth.

Interestingly, finding fraudulent trials is not that hard to do, especially in the Middle East, a region well known for publishing shady research. Keep that in mind.

Hill's game-changing paper was published on July 6, 2021. Eight days later, one of the largest and most positive trials included in his meta-analysis was retracted from the preprint server it had been posted on. Lead-authored by Dr. Ahmed Elgazzar, a Professor Emeritus at the University of Benha in Egypt, the paper found ivermectin to be highly successful in reducing hospitalization and death in Covid patients. Further, in another arm of the trial, he showed that ivermectin was highly effective in preventing Covid infections as well. This Elgazzar paper had been posted on the preprint server eight months prior, mind you, and was suddenly retracted a week after Andy's analysis was published. Purely coincidental or highly suspicious? You know what I think.

The very day after the Elgazzar paper was retracted, a man named Jack Lawrence from the UK, a master's degree student and the founder of Grftr News (a site allegedly dedicated to "writing about and researching misinformation and bad science") published an article titled, "Why was a major study on ivermectin for Covid-19 just retracted?" His 3,000-word answer to that question contained the words fraud, fabrication, plagiarism, and ethical breaches, to name a few.

"The study had formed a crucial piece of evidence in the pro-ivermectin case and *its removal largely destroys (the current) scientific case for using the drug in COVID-19 care* [emphasis mine]," Lawrence explained in his hit piece.[4]

The Guardian—which had unilaterally ignored ivermectin until this point but was hot to cover it now that it had been "debunked"—ran a splashy story on Lawrence's discovery. My favorite line shows just how well-trained Lawrence was in the dark arts of public relations campaigns:

> Lawrence said what started out as a simple university assignment had led to a comprehensive investigation into an apparent scientific fraud at a time when "there is a whole ivermectin hype . . . *dominated by a mix of right-wing figures, anti-vaxxers and outright conspiracists.*[5]

Overnight, the media narrative was carved into stone: "Ivermectin doesn't work, and the studies are all fraudulent."

The shills came out in full force on Twitter in particular. Leading the pack were a pair of gentlemen (I use the term loosely) by the names of Gideon Meyerowitz-Katz (@GidMK) and Kyle Sheldrick (@K_Sheldrick). They were two mostly unpublished nobodies in the world of science and piddling characters on social media who would go on to publish papers with decades-long, previously celebrated researchers like Andy Hill and Ed Mills, another heavily Pharma-conflicted ivermectin researcher.

Meyerowitz-Katz and Sheldrick aren't the only sellouts shilling for Pharma on social media, but they stand out for their concerted and repeat offenses. Sheldrick would later go beyond ivermectin, taking to Twitter with a ludicrous statistical analysis of Paul Marik's original vitamin C trial, and claiming to "prove" that it was fraudulent. I am not making this up. The most published and one of the most respected physicians in the history of critical care medicine being publicly accused of fraud by a nobody spouting wildly erroneous statistical analyses.

Sheldrick went even further, writing a letter to the journal *Chest*, requesting that they retract Paul's IVC paper.

Don't think for a minute we were going to let that one slide. A group of colleagues rebutted Sheldrick's shoddy analysis in a twenty-page treatise which we sent to *Chest*. *Chest* responded by launching an extensive investigation into Paul's study by asking Sentara Norfolk General Hospital, where Paul worked, for the source data. It took a year, but Paul finally received the response from *Chest* that they had found no ethical concerns and would not be retracting the paper. He was in tears as he shared the news with me. I honestly had not realized how hard this whole ordeal had been on Paul. If that paper had been retracted, it would have tarnished his entire prolific career of publications. I had simply considered it just another assault in a long list of attacks on me and Paul, but a retraction of that nature would have been crushing.

Long before we learned of *Chest's* decision, we threatened Sheldrick with a lawsuit for defamation. Wisely, he caved and quickly agreed to issue a public retraction of his accusations. But it's important to include the backstory here, as Sheldrick's actions clearly illustrate his character and credibility—as well as those of whoever he is or was working for. It's also worth noting that *harassing scientists who produce inconvenient science* is a disinformation tactic called "the blitz." Its main aim is to destroy the credibility of those producing inconvenient science so that their science is not believed. Sheldrick tried to blitz Paul for his vitamin C study—and failed spectacularly.

It was a much-needed win for the FLCCC.

As for Meyerowitz-Katz, the record of his anti-ivermectin statements and actions is almost incalculable, but the c19early group has done a fantastic job documenting the lot. His "disinformation record"[6] reads like a career criminal's rap sheet. Alexandros (Alex) Marinos, PhD, a technology entrepreneur, researcher, and hero in the ivermectin corruption detection world, also wrote a powerful analysis of all of both Meyerowitz-Katz's and Sheldrick's indefensible actions and the public statements they used in trying to create this narrative.[7]

For instance, Meyerowitz-Katz and Sheldrick claimed to have analyzed a random set of ivermectin RCTs and found them to be fraudulent or have other "major issues." Meyerowitz-Katz promised to release this momentous analysis . . . almost a year and a half ago, as of this writing. To date, their information has never been shared. Which is odd, because Meyerowitz-Katz once publicly tweeted, "When people running clinical trials refuse to share anonymized raw data, it is a giant red flag without a very good excuse."[8] You don't say.

To be clear, the Elgazzar paper was not without issue. For starters, the incentives to publish positive studies that get attention are considerable. For this reason, fraudulent studies are therefore not uncommon (one researcher quoted in *The BMJ* estimates that about 20 percent of trials are false[9]). When the story broke, Tess Lawrie and I both wrote to Elgazzar; he claimed that the database of patient data that was being circulated and attacked as fraudulent was "not the source data for the trial." We never heard from him again, despite asking him repeatedly to share his *real* source data. It is my belief that either he is a fraud, or he and his university got paid to be the fall guys (I lean towards the latter).

Either way, it doesn't matter. Every single medicine studied will have a percentage of trials that are fraudulent. This is in no way unique to ivermectin, yet the narrative consistently fails to recognize this fact. Instead, they attempted to paint the entire body of evidence for ivermectin as uniquely fraudulent. Even if half the studies on ivermectin were bogus, you'd still have a massive evidence base of efficacy, especially when you include the epidemiologic data following ivermectin distribution campaigns, or from my and many thousands of other doctors' treatment experiences across the world.

The c19early group calculated how many studies you would have to remove in order to "disappear" the signal of efficacy. They found that you would have to omit sixty of the ninety-five studies to come to a non-statistically significant conclusion (note that the ninety-five studies already exclude Elgazzar's study in prevention and treatment). They even tried to play along with Meyerowitz-Katz and Sheldrick by re-analyzing the data after excluding all the studies this group of ivermectin assassins publicly called into question. Guess what happened? *The signal of efficacy got even stronger.*

But this destructive PR campaign by the media and academia had been launched. The narrative was fixed: *ivermectin science can't be trusted.* They published articles in *Forbes,* the *Wall Street Journal, BBC News,* even *The BMJ (British Medical Journal),* smearing ivermectin six ways to Sunday. The pseudo-experts were quoted repeatedly, saying "the ivermectin literature is full of fraud" (Meyerowitz-Katz), and, "If you asked me twelve months ago, I would never have guessed a fraction of this rate of fraud" (Sheldrick).

During the pandemic, Alex Marinos, the Covid fraud hunter, dedicated himself to the study of clinical research fraud. With his wife, Eva Tallaksen, in 2021 he established BetterSkeptics, a transparent, open-source platform for crowdsourcing Covid facts. The goal was to shed light on the limitations of the so-called fact-checking that was patently biased against anyone like me. Marinos repeatedly challenged Meyerowitz-Katz and company, documenting their actions (and lack of them) on Twitter and in countless Substack posts.[10] One commenter on a recent post offered this summary of the scandal: "Fraud is almost always preceded by a motive which, if successful, gives the fraudulent party a benefit; and generally weighed against the risk of being caught. (In this environment), a positive Ivermectin study provides the fraudulent researcher with zero money as ivermectin is a generic, and two - character assassination, slander, defunding of research, firing from a position. I am waiting to hear someone provide even the slightest motivation for submitting a fraudulently positive Ivermectin paper - you still get 1 and 2, plus you're a fraud."

(Alex also uncovered an article Meyerowitz-Katz had authored in 2019 defending Monsanto's cancer-causing weed eater Roundup.[11] Nothing to see here!)

When Tess, Paul, and I learned of the allegedly fraudulent Elgazzar paper, our ivermectin review had already been published in the *American Journal of Therapeutics*. The editor asked us to re-analyze our data without this study and we quickly did so. We found that the treatment effect was weakened slightly but did not alter the conclusion.

What we did in that moment was an act of medical integrity; what Andy Hill did next was, in my opinion, an act of calculated degeneracy that shocked me because *I knew that Andy knew ivermectin was lifesaving.* First, rather than simply revising his paper like we had, he decided to self-retract his highly positive paper. Then he not only removed Elgazzar's trial from the analysis (fair enough) but also removed any trials which he had graded as "high risk of bias." This isn't how meta-analyses work. (For one thing, it's standard procedure to include a mix of quality gradings in analyses; secondly, there is absolutely no evidence to support the pervasive belief that high quality and low quality trials reach different conclusions.)[12] If this wasn't bad enough—and trust me, it was—Andy decided to invent two new categories of trials, "potentially fraudulent studies," and "studies with some concerns," *and then proceeded to remove the trials which fell into his made-up categories.*

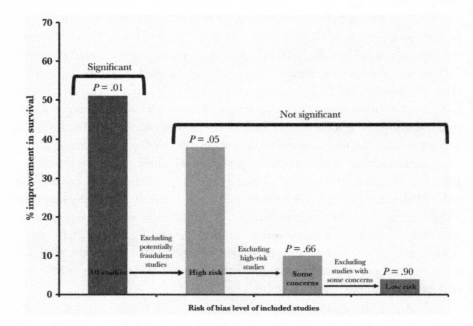

I could find no definition for these categories in the paper. Andy's new "analysis" was also published in a scathing editorial in the *New England Journal of Medicine*, only the top medical journal in the world—which apparently had zero qualms about including his entirely made-up grading categories.

The chart that accompanied his analysis looked like something a first grader would create to try to convince his parents to increase his allowance. And the *NEJM* published it! I'd laugh if I weren't extremely busy trying not to punch something.

By arbitrarily categorizing the studies in this way, Andy was able to whittle down the alleged evidence base to just four trials to estimate mortality. It was the exact same tactic that the WHO had used in creating their ivermectin non-recommendation. Although all the benefits reported in his original paper were still found, the data was insufficient to achieve statistical significance with regard to preventing death. He went public with the brazen lie that his analysis showed ivermectin no longer had an impact on mortality.

The vast majority of the world either didn't care, wasn't paying attention, or simply did not understand how evidence-based black magic is practiced.

The FLCCC cared. Tess cared. I cared to the point of absolutely losing it one night, after Andy posted yet another infuriating tweet. "More evidence still needed from new trials of ivermectin," he wrote. "Current trials database does not provide clear conclusions." He included a link citing the lack of "high-quality" RCTs.[13]

I probably shouldn't have been on Twitter at 3:41 a.m., but I was. And I was pissed.

So, I fired back. With an expletive. Two of them, in fact. Whoops.

I'm not proud of that tweet but I also don't regret it. However, I did delete it the next day on the suggestion of Bobby Kennedy Jr., a man whose wisdom, professionalism, and advocacy I deeply admire. I wish I could be more like him. Maybe someday. But certainly not that day.

If you recall, when I first met Andy Hill, he told me he was working on a team funded by Unitaid, whose mission was to find and study repurposed drugs for use in Covid. I no longer believe that was the true mission of that team. I am now convinced their real mission was to find evidence of efficacy of repurposed drugs *so that BMGF and Big Pharma could initiate their disinformation campaigns swiftly and mercilessly.*

Pierre Kory, MD MPA ⋯
@PierreKory

Replying to @DrAndrewHill

Andy, STOP. Seriously, you are causing untold deaths man. For another WHO paycheck in the future? WTF. Me and Tess have both blown up our careers because history demanded it. Your fake cautiousness is the saddest shit I've ever seen. Fuck you. And that is putting it mildly.

3:41 AM · Jul 31, 2021 · Twitter for iPhone

413 Retweets   **65** Quote Tweets   **1,405** Likes

♡   ⇄   ♥   ⬆

I don't know who's been paying Andy's bills since the fraud narrative campaign began and his Rainwater funding ran out; what I do know is that he's now an official member of the above-mentioned group of ragtag bloggers and social media small-timers who ignited and fueled the anti-ivermectin inferno. They work and publish together in tandem now. Like in real journals, not just on social media. Any time you see a corporate-controlled media article attacking ivermectin, you can bet it has one (or all) of their damaging quotes in it.

Andy Hill, Andrew Owen, Dominique Costagiola, and their Pharma and BMGF sponsors destroyed ivermectin at the global level. Fauci and his minions, using the captured US media, eliminated it as an option in the US. Every advanced health economy had a Fauci-like figure who did the same. Despite my hatred of these crimes, I agree with Phil Harper that what happened is far bigger than any two or three individuals. At the heart of this tragedy is a system that has immense power to influence behavior, and the vast majority of people are powerless against it. Andy, Andrew, and Dominique weren't special. If it weren't those three, it would have been three others.

In the end, I believe Andrew Owen directly profited from his complicity with that system, while I think Andy Hill eventually submitted due to cowardice at the prospect of having his livelihood and future career prospects destroyed. I guess it's the same thing. Either way, defy BMGF and you are done in international public health. *Done.* Maybe I'm naive, but I don't believe Andy knew he was doing opposition research at the beginning of his investigation into ivermectin. But his later actions leave no doubt in my mind as to what ultimately happened.

They got him.

They always do.

CHAPTER TWENTY-FIVE

Counterfeit Trials—The Big Six

While all the manipulation and distortion of the massive evidence base of ivermectin's efficacy was going on at the WHO, Unitaid, NIH, and BMGF, this "other side" was also executing their own disinformation tactics. I maintain that their primary and most impactful play was "the fake," which can be run four different ways in regard to the science around a repurposed drug:

1. **Manufacture counterfeit science** by commissioning studies with flawed methodologies biased toward predetermined results, or "designing trials to fail"
2. **Selectively publish negative results** in scientific journals
3. **Block publication of positive results** in scientific journals
4. **Plant ghostwritten articles** in scientific journals

The most devastating maneuver was the first one, *commissioning studies biased towards predetermined results*. Their subsequent "large," "high quality," "rigorous" trials fueled the media and academic narratives that rippled around the world. To recount and document all of the deceptions embedded within those trials would obliterate my publisher's imposed word limit on this book, so I will present merely the highlights of what I believe to be the most damaging actions that were executed.

As of this writing, there have been six major ivermectin trials in Covid that have been published in high-impact medical journals. Each of the "Big Six" somehow found that ivermectin was not useful in preventing or treating Covid. Each one's publication led to a media firestorm of headlines blaring around the world that "ivermectin doesn't work," accompanied by the falsely definitive statement, "says a recent, large, high-quality RCT."

The Big Six trials have a frightening number of disturbing characteristics in common.[1] First and foremost, each was designed and conducted by expert trialists with massive conflicts of interest with Pharma or BMGF (which are sort of the same thing, given that both share the same goal to vaccinate the entire world, and both have bottomless piles of money to influence anyone and anything). Many of the investigators worked indirectly or directly for BMGF but most worked or received money from . . . wait for it . . . *competitors to ivermectin*, either vaccine manufacturers like Merck (molnupiravir) and Pfizer (Paxlovid), or companies producing monoclonal antibodies. The principal investigator of both NIH ivermectin trials has several notable conflicts of interest, including owning stock in a competitor to ivermectin. In fact, I reviewed every one of the thirty-seven trials studying ivermectin as an early treatment, and of those thirty-seven, only one trial outside the Big Six had a conflict of interest. All the other ivermectin trials, where the authors declared "no financial conflicts of interest," consistently showed benefits.

Anyway, each of these six large, "high quality" trials was not only performed by study investigators drowning in Big Pharma conflicts, but was, unsurprisingly, brazenly fraudulent in their design and conduct. And while every single trial showed ivermectin to be superior to placebo, none of the benefits reached the hallowed level of "statistical significance." For a drug this effective, it is not surprising that, despite the insane amount of trial manipulations I will detail below, they could not disappear the benefits completely. But, in the age of modern "evidence-based mania," it's enough just to lessen the differences to come under the level of statistical significance. And that is what they did. Over and over again.

One of the trialists' main weapons, which escaped the attention of the entire world, is that most of the studies were conducted in South American countries where ivermectin was not only available over the counter, but its efficacy was increasingly recognized by word-of-mouth, through social media, and even via local media and government communications. There is abundant evidence[2] that the control groups in South America were not only taking ivermectin but also that the Pharma-conflicted study investigators were taking little to no action to identify those taking ivermectin and to exclude them from participation. Which makes it very, very hard to prove that *ivermectin is more effective than ivermectin.*

Further, with both groups getting ivermectin, there were very few hospitalizations or deaths, the result of which is what is called an "underpowered trial." By definition, when a study is underpowered, most differences between treatment groups will not reach statistical significance. It's a double whammy. With rare exception, these limitations were not mentioned in the trial summaries and certainly not in the conclusions. Yet, each time they were published, the uncritical news media went wild screaming, *"See? Ivermectin doesn't work!"*

A more accurate headline would have been, "Ivermectin was found to be no more effective than ivermectin."

Complaints about the innumerable flaws in study design and conduct fell on deaf ears, found zero media traction (shocker), and were routinely dismissed as the whining of sore losers by social media mavens when we would point them out. Yet the same nefarious design tactics kept appearing and reappearing. And again, if you weren't in the routine habit of reading and analyzing clinical trials, you wouldn't be expected to pick up on the fraud . . . which is precisely what the crooked "researchers" were counting on.

It's almost absurdly comical to see how easy it was to hijack the entire narrative when you understand their tactics. Here's my ivermectin crime scene breakdown, focusing on the Big Six brazenly illicit trials: Lopez-Medina, I-Tech, TOGETHER, COVID-OUT, ACTIV-6 (400), and ACTIV-6 (600).

Tactic #1: Not excluding subjects taking ivermectin from the control/placebo groups and/or *giving the placebo group ivermectin*. In the Lopez-Medina trial, the latter is actually what happened; in the paper, it was described as a "pharmacist's error." I suspect that the only error the pharmacist made was coming to work with integrity. I and many colleagues hypothesize that this was an honest pharmacist, who after discovering the placebo and ivermectin bottles were "mislabeled," likely threatened to go public. So they said "whoops," and allowed him or her to "fix" the problem and they reported it as an honest error in the paper. Alex Marinos's alternate hypothesis is that, during the three-week period the placebo group was supposedly getting ivermectin, the data for the real ivermectin group was particularly strong so they invented this excuse so they could throw out that stretch of data. Either way, this "error" stinks to high heaven.

Tactic #2: Enrolling only mildly ill, young, or otherwise healthy patients with a primary endpoint of measuring differences in hospitalization.

Shocker: so few went to the hospital that it underpowered the trial (these researchers are malevolent, but not stupid).

Tactic #3: Administering as low of a dose as possible.[3] This is like giving a severely dehydrated person a single sip of water and then declaring, "water is ineffective at improving hydration."

Tactic #4: Administering ivermectin on an empty stomach despite knowing full well that concentrations are higher with a meal and that higher concentrations lead to more potent efficacy. (For these reasons, the FLCCC has long recommended taking ivermectin with a meal. It's not rocket science.)

Tactic #5: Placing arbitrary, invented, unprecedented weight limits to dosing such that the highest risk patients (obese) were under-dosed. To wit, the only trials I have ever been able to find with such weight dosing limits imposed were . . . the Big Six. It's incredibly telling that a weight limit for dosing ivermectin had never been described before appearing in these trials.

Tactic #6: Publishing trial data after an unnecessarily long delay. The repurposed drug trials, especially those studying ivermectin, were the only ones in the pandemic to take extraordinarily long times to publish after the trials finished. Contrast these delays with the immediate issuing of study results by press release employed by Pfizer with Paxlovid, Merck with molnupiravir, and Gilead with remdesivir. Apparently, manipulating data to hide efficacy takes a lot of time. No explanations were ever given for these delays by any group of authors of these trials.

Despite my focus on the Big Six trials conducted by Pharma-controlled researchers thus far, technically it should be the "Big Seven." So why am I leaving one out? Because Oxford's PRINCIPLE trial results have not been released yet, despite having completed enrollment in July of 2022. It's now April of 2023. *Nine months later.* Curious, no? It's literally the biggest trial of them all, and that trial has gone quiet. Despite the fact PRINCIPLE employed one of the most biased study designs of the Big Six (I mean Seven), I suspect the principal investigator Professor Butler was shocked to find that the study data revealed a massive efficacy (ivermectin is that good). He and his colleagues are either still figuring out how to best distort or manipulate the data to show ivermectin is ineffective, or they don't want to even try to do so publicly because it would incriminate them. But we are onto them; unbiased and unconflicted researchers like Alex Marinos, Tess Lawrie, Phil Harper, Jackie Stone, David Wiseman, Colleen Aldous, and many others are watching closely. I suspect they know this, and that they're just laying low, hoping no one notices they are not publishing the results. We notice. Unfortunately, the media doesn't care about what can only be viewed as scientific misconduct at this point. Project Veritas or James O'Keefe, if you're reading this, we need you.

Tactic #7: Enrolling patients many days into the disease, after their trajectory was already set.[4] This study design feature was the opposite of what Pfizer did in their Paxlovid trial where all subjects were treated within three days of onset of symptoms. It also was the opposite of what happened in a recent molnupiravir trial, where they enrolled 25,000 patients *a median of two days from first symptoms*. Please read that again. That is actually a very impressive feat, and one that I have never heard of being achieved in any RCT of an acute illness with such a size. Conversely, all the trials of ivermectin allowed up to seven or more days to enroll in the study, and the most ludicrous was again Oxford University's PRINCIPLE trial, which allowed enrollment up to fourteen days after the onset of symptoms (which is insanity). Also note that both Merck's molnupiravir trial and the PRINCIPLE trial have the same chief investigator:

	Molnupiravir	Ivermectin
Trial	PANORAMIC	PRINCIPLE
Chief investigator	Prof. Chris Butler	Prof. Chris Butler
Randomization delay	Median 2 days, ≤5 days from onset	≤14 days from onset (median unknown)
Population	50+ or 18+ w/comorbidities	18+ (mid-trial change, prev. 18+ w/dyspnea or comorbidity, 65+)
Treatment	5 days, 2x per day	3 days, 1x per day, dosage below real-world protocols and recent trials
Patients randomized	25,783	est. 4,500
Enrollment period	Dec 8, 2021 - Apr 27, 2022	May 12, 2021 - Jul 8, 2022 (est.)
Cost	$707	<$1 (off patent)
Merck profit	$5.4B sales to June 30, 2022 (2021, 2022). Estimated $17.74 to produce.	~$0 (potential, unlikely competitive with low cost manafacturers)
Mutagenic	Yes	No
Design better for showing efficacy		
Design worse for showing efficacy		

Interesting sidenote: Around this time, a British musician named Michael Beasley wrote to me from the UK. He was sick with Covid and hoping to get ivermectin. I suggested he reach out to the PRINCIPLE trialists at Oxford in the hopes he would be enrolled in the trial and possibly get ivermectin instead of placebo. He was accepted and sent a three-day course of pills to take. He reported to me that he was symptom-free in twenty-four hours and virus-free in forty-eight. He later wrote to me that he was able to source his own ivermectin from India and that he used it with equal success the following year. I'm sure it was just the placebo effect both times.

Tactic #8: Changing the rules in the middle of the game. In most or all of the Big Six trials, critical components (including registered study protocols dictating which patients to include or exclude, how to perform the randomization, how to analyze the data, and what dose to be used) were changed *during the actual trial*. This is again supposed to be a never event in research trial conduct. And not only did they pull these moves multiple times with impunity, but they had no problem getting past peer review in the top medical journals in the world. Had any of us tried to pull this stuff in an ivermectin trial, it would have invalidated the trial and it would not have been published in any journal.

Tactic #9: Refusing to share the individual patient data (IPD) or source data of the trial. Requiring the source data of clinical trials to be shared publicly after publication has been a goal of many physicians and researchers eager to "clean up" the corruption in science over the past few decades. Remember our "friend" Gideon Meyerowitz-Katz, who stated that *anytime study authors do not release the source* (IPD) *data, this should be considered a major red flag.* That is the one thing Gideon and I agree on. Not one of those Big Six trials have released their IPD data publicly. The most troubling instance was the TOGETHER trial of Ed Mills. In their protocol and in their paper, they stated repeatedly that the IPD data would be made publicly available, first at the "termination of the study" and then later, they switched it to "after publication." Dozens of researchers have asked the investigators repeatedly for the IPD data since the publication of that trial. Crickets.

The most brazen example of fraud came from our very own, esteemed National Institutes of Health (NIH). Two years into the pandemic, they finally decided to fund a "proper" trial, to see, once and for all, whether ivermectin effectively treats Covid. They pulled every shady study design trick mentioned above. Every. Single. One. Blatantly Pharma-conflicted investigators, the lowest weight limit to dosing yet, an impossibly short duration of treatment, delayed delivery of ivermectin because they shuttered the study on weekends, insanely late enrollment, studying only mildly ill patients, you name it. The irony is . . . *they failed to find ivermectin inferior.* In fact, they found a highly statistically significant benefit in their primary outcome (just like what I suspect happened in Oxford's PRINCIPLE trial). So what did they do? They were forced to bust a new move, one which you are *never* supposed to do in clinical trials: *They changed the study endpoint.[5] In the middle of the trial.* An unequivocal scientific never. (Except if you are Pharma; Gilead pulled this trick in their remdesivir trial and it still sailed to publication in the top journal in the world).

Here's how it played out:

Recall that after Paul, Andy, and I presented to the NIH, they updated their recommendation to a neutral one, but still avoided even the weakest recommendation available to them and for which the data far exceeded the necessary criteria. One of the authors of this recommendation was Susanna Naggie of Duke University's Clinical Research Institute (DCRI). Two months after she voted for this neutral recommendation, the NIH announced a $155 million award (NOT-TR-21–024) for the "funding of ACTIV-6 to study the use of repurposed drugs in Covid."

It all went to Susanna Naggie's team at DCRI.

A weak recommendation would have removed the possibility of doing a placebo-controlled study with ivermectin alone. Remember the need for "clinical equipoise" in order to conduct placebo-controlled trials? Further, what US citizen would agree to enter a placebo-controlled trial for a recommended treatment? There goes the $155 million grant! According to the NIH, DCRI was "the only institution in the country that met the criteria" for the award. I wonder what those criteria could have been.

Naggie's conflicts of interest with ivermectin competitors include but are not limited to accepting personal fees from Gilead (remdesivir), Abbvie (monoclonal antibodies), Pardes Biosciences, Personal Health Insights, and Bristol-Myers Squibb. She also holds stock options in Vir Biotechnology, which developed a monoclonal antibody called Sotrovimab in conjunction with GlaxoSmithKline. Sotrovimab was being developed for use against Omicron and its subvariants, which means Naggie owned stock in a therapy whose market would be obliterated if ivermectin was proven effective.

Further, the original endpoint of the study was "time to clinical recovery at fourteen days." In the middle of the trial, they switched the endpoint to twenty-eight days. Tellingly, the benefits in the ivermectin arm were statistically significant at Day 7 as well as at Day 14. But when they moved the endpoint to twenty-eight days, poof! The statistical significance disappeared.

The abstract conclusion reads: "Among outpatients with mild to moderate COVID-19, treatment with ivermectin, compared with placebo, did not significantly improve time to recovery. *These findings do not support the use of ivermectin in patients with mild to moderate COVID-19.*"

These trial manipulations are not the only ones we identified, but they are the most heinous, blatant, and consistent ones. (If you want to see the rap sheet of crimes meticulously compiled on the worst of the Big Six, go to c19early.org.[6] Reading those gives me the howling fantods. If you know, you know.) Despite these repeated actions, in their papers, the study authors consistently and often definitively concluded that ivermectin had no efficacy. Sometimes they went even further, issuing blatantly idiotic policy-making statements such as "these findings do not support the use of ivermectin for treatment of mild Covid-19." Policy statements, in my career, have *never* appeared as part of a study's conclusion. Well, until Andy Hill's manipulated early preprint version. Researchers are not policy makers. Tragically, only a handful of doctors read beyond the abstract conclusions. Hell, most don't even read medical journals. From what it looked like, most were learning about Covid, ivermectin, and the vaccines through the media. (Even CDC director Dr. Rochelle Walensky admitted to getting her pandemic information from CNN.)[7] The narrative was fixed, and Pharma knew it.

Here's the excruciating part: The fraud that was expertly baked into these "rigorous" studies from the beginning would form the foundation of scientific lies that to this day supports the entire edifice of the relentless, single-track media messaging against ivermectin. It was a calculated construct and one that could only be built by shameless criminals.

CHAPTER TWENTY-SIX

The Editorial Mafia

The world's highest-impact medical journals are the *New England Journal of Medicine (NEJM)*, the *Journal of the American Medical Association (JAMA)*, *The Lancet*, the *British Medical Journal (BMJ)*, and the *Annals of Internal Medicine*. (Cochrane Library is considered the top database of systematic reviews and meta-analyses.) The FLCCC's Flavio Cadegiani has dubbed the editors of this group "the Editorial Mafia" for reasons which will become apparent. Although the "high impact" designation is a numerical one determined by the number of citations each receives annually, I interpret the term differently since Covid. High impact to me now means that when a study is published in those journals, *it immediately drives news headlines*. Second and third tier publications (also ranked based on how often they are cited by other journals) never make news. Pharma knows this.

The high-impact journals alert the press to any new or important study findings, whether positive or negative, *prior to publication*. Science reporters (who with rare exceptions seem to have abandoned science in favor of running PR campaigns for Pharma), get to ask the study investigators questions about the study design, conduct, or conclusions and thus they can write up newsworthy headlines and articles which appear on the same day as the study's publication.

All of the Big Six trials I summarized in the last chapter were published in high-impact medical journals. Each generated massive negative PR campaigns against ivermectin in the world's newspapers, TV, and radio stations. Each time the trial papers went through peer review at those journals, the incalculable design flaws and conduct shenanigans were never mentioned in the paper's "limitations" section, which is a standard part of every trial manuscript. In other words, the peer reviewers of those trials allowed the authors to publish without forcing them to highlight the tremendous, glaring limitations or to include the critical data needed to fully assess ivermectin's impacts.

It was as if there was an Emergency Ban on scientific integrity around the globe.

In theory, peer review exists to ensure that the methods of data collection, analysis, and interpretation of a study are valid and support the study's conclusions. You discover something potentially phenomenal; you present this discovery to equally qualified experts, and they all either agree that it is phenomenal—or poke massive, un-survivable holes in it. Peer review was designed to be a checks and balances system used to maintain quality and integrity in scientific research. That is, of course, assuming that everyone involved acts with integrity—which is the exact opposite of what was happening both blatantly and rampantly. But by this point, "peer review" had become synonymous with "fact checked." It's akin to continuing to use the argument "but Snopes said it's false," after the so-called fact-checking site was caught plagiarizing dozens of articles from mainstream media outlets. I believe the adage is, *consider the source.*

I was part of a large network of ivermectin researchers who were analyzing these "designed to fail" studies. When the first of the Big Six was published in *JAMA* (Lopez-Medina), one hundred scientists wrote an open letter to the journal calling for its retraction. We were ignored. The TOGETHER trial, partly and secretly funded by BMGF (Alex Marinos busted the TOGETHER trial for removing evidence that BMGF had helped fund the ivermectin portion of the trial), ignited an even more intense uproar due to even more brazen fraud that we uncovered.[1]

Alex, Phil, and the c19early.org group call the TOGETHER trial "The Trial of Impossible Numbers" because, in their multiple public presentations and publications of the data from the ivermectin, fluvoxamine, and metformin arms of that trial, the data are impossible to reconcile.[2,3] Further, numerous pieces of critical data were left out of the study publications.

What's crazy about this trial is that it also had an arm which studied a Pharma drug, Pegylated Interferon Lambda, made by Eiger Pharmaceuticals. Although TOGETHER found the drug to be effective, when Eiger attempted to apply for an EUA, *the FDA denied the request for a pre-EUA meeting*, citing concerns about the conduct of the TOGETHER study.[4]

Despite all of this, the TOGETHER trial was published in the *New England Journal of Medicine*, the top medical journal in the world, which exploded a media circus across the world against ivermectin. Again, our requests for retraction, angry letters to the editor, and repeated appeals for the IPD data were ignored.

Dr. Marcia Angell, the long-time editor-in-chief of the *NEJM*, resigned over twenty years ago in June 2000, after two decades in the post. She left because of what she described as the rising and indefensible influence being exerted by Pharma at the prestigious journal and its powerful affiliate societies. *Back then.*

Dr. Angell wrote a remarkable book about the widespread corruption she witnessed. In *The Truth About Drug Companies: How They Deceive Us and What to Do About It,* she wrote:

> Now primarily a marketing machine to sell drugs of dubious benefit, big Pharma uses its wealth and power to co-opt every institution that might stand in its way, including the US Congress, the FDA, academic medical centers and the medical profession itself.[5]

The above is exactly why I call our country the United States of Pharma. One reason for this is that between two-thirds and three-quarters of Big Pharma's global profits are made in the USA. This is largely due to the fact that drug prices are not regulated in this country, leading to comical, astronomical prices insurers happily pay. I have been saying for nearly three years that our federal health agencies are (and have long been) in a state of total regulatory capture, defined by Wikipedia as, "a form of corruption of authority that occurs when a policymaker or regulator is co-opted *to serve the commercial interests of an industry.*"

In her book, Dr. Angell brilliantly and unapologetically summarized the sorry state of the medical industrial complex:

> It is simply no longer possible to believe much of the clinical research that is published, or to rely on the judgment of trusted physicians or authoritative medical guidelines. I take no pleasure in this conclusion, which I reached slowly and reluctantly over my two decades as an editor of the *New England Journal of Medicine*.

Further, Angell had no qualms pointing out something that's becoming clearer by the year: sick care is big, big business. In fact, Angell wrote, "*In*

2003, the profits of the top 10 big Pharma exceeded that of the cumulative profits of the other 490 Fortune 500 Companies."

Let those numbers sink in for a moment.

This deep-seated corruption is hardly new. The late Dr. Arnold Relman, another former editor-in-chief of the *NEJM*, said this in 2002: *"The medical profession is being bought by the pharmaceutical industry, not only in terms of the practice of medicine, but also in terms of teaching and research. The academic institutions of this country are allowing themselves to be the paid agents of the pharmaceutical industry. I think it's disgraceful."*

The most damaging aspect of Pharma's complete control of the high-impact journals is that most doctors (including the old me) have no idea that it is occurring, and even if they do, they could never truly comprehend the extent or the direct, depraved consequences of it. The crooked Editorial Mafia know exactly what they can and should publish in those journals to benefit Pharma—and what they cannot. Since most doctors are in the dark about this practice, they place an immense amount of trust and confidence in these "esteemed" journals. Further, medical practitioners are unaware of the science that is *not* published in those journals, namely the research supporting interventions that effectively prevent or treat disease but offer little profit to the pharmaceutical industry. In the recent past, millions of people have "woken up" to this pervasive corruption as the result of the innumerable policies issued that were divorced from supportive or rational data that many of us were compiling from non-high impact journal sources. As a result of all this, many people are now terrified of going to any doctor, viewing them for what they unknowingly have become: Pharma-controlled automatons arrogantly and unknowingly parroting fraudulent "science" and unsound treatment and vaccination guidelines.

Dr. Aseem Malhotra is a prominent British cardiologist and Covid truth-teller who was initially and vocally pro-vaccine, but famously switched teams after he concluded that his father died as a result of the Covid vaccine. In a recent interview, the reporter asked, "How worried should we really be? It's quite easy to paint a picture of nefarious goings-on of Big Pharma as a huge shadowy industry who doesn't care about our health and only cares about profit margins . . . but is it really that bad?"

Malhotra couldn't answer quickly enough. "It's very bad, actually," he insisted. "It comes back to [the fact that] the legal obligation of the drug industry, people need to understand, is to produce profit for shareholders.

They have no legal obligation to give you the best treatment. The real scandal . . . is that doctors, institutions, and medical journals collude with industry for financial gain and the regulators fail to prevent misconduct by industry. We have a wealth of evidence of fraud that's been committed by the pharmaceutical industry over the years but for them it's the cost of business."

Let me give you an analogy: Imagine that the fine for robbing a bank was $1 and a disappointed slap on the wrist. Now imagine that the average bank robber walked away with ten thousand dollars per heist. If you got caught (and sometimes you wouldn't!), you'd get to keep the ten grand—but you'd have to pay a whopping one-dollar fine. At the risk of underestimating humanity, I'm pretty sure ninety-nine percent of the world's population would quit their jobs and become bank robbers.

But back to the Big Six. Every one of these trials is burned in my memory because on each of their publication days, I woke up to a barrage of frantic FLCCC emails, texts, and phone calls. The media firestorm was fierce, with "reporters" arrogantly celebrating their belief that we had "gotten it wrong."

Each day was a PR crisis, with our team in a mad scramble to defend our credibility and treatment recommendations. I hated each and every one of those days. Hundreds of pompous doctors tweeting and posting some version of "See? I told you it didn't work!" So-called journalists were cranking out hit pieces at Mach speed, calling us "advocates" at best and "fringe anti-vaxxers" at worst. It was relentless and exhausting.

Watching those trials get published and then heralded as "large, high quality, rigorous trials" celebrated across academia and the media was anguishing. Hundreds of thousands of doctors across the world who were using ivermectin and knew it was efficacious had to witness the same thing. We watched helplessly as ivermectin either became outlawed (Australia), removed from hospital formularies (the entire US), and/or was restricted in use by arrogant pharmacists. All while case and death counts rose in massive waves and as a result of counterfeit or mis-interpreted science.

There was something predictably sinister about the publication dates of the Big Six. March, 2021; July, 2021; February, 2022; May, 2022; October, 2022; February 2023. It seemed to me, from my front row seat, that these studies and their mass media campaigns were spaced out in strategically timed intervals. It was like a cannon firing every few months, with each propaganda round sounding louder and traveling farther, until everyone in the world thought the same thing: *ivermectin doesn't work.*

This hypothesis might explain why the results of the TOGETHER trial (the most fraudulent and impactful of them all) were not published until *nine months* after the study was terminated. I think they held onto it in order to time its impact for whenever there was a "dry spell" of bad news against ivermectin. Or maybe it just took that long to distort and manipulate the data and protocol sufficiently.

If there's one thing I can say Pharma is good at, it's conducting disinformation campaigns.

CHAPTER TWENTY-SEVEN

The Journal Retractions of Positive Ivermectin Studies

Another element of "the fake" is to censor positive reports; in the case of Covid, they specifically went after positive reports of repurposed drugs. While Big Pharma and BMGF were paying their trialists to conduct and publish counterfeit science, they not only pressured journal editors to retract positive peer-reviewed and published papers of ivermectin, but also induced them to reject any submissions of ivermectin trials with a positive result, no matter how high quality their design. Remember, this was a war on ivermectin, and the generals leading the opposition were the Editorial Mafia.

The first retraction victim was the FLCCC's wickedly positive review paper with me as first author. Despite passing three rounds of rigorous peer review by three senior scientists at the FDA and NIH and a fourth independent expert ICU doctor, *The Frontiers of Pharmacology* sat on our paper for weeks. This was the winter of 2020–21 and hundreds of thousands of people were dying around the world. We knew something was really wrong. I finally threatened the journal's editorial representative that we were going to go public with an accusation of scientific misconduct. A global pandemic with an obvious solution and they were slow-walking it to publication? It was an online journal!

Dr. Robert Malone, an outspoken Covid war hero, was the special guest editor of the issue I had submitted to; the entire issue was to be dedicated to research into repurposed drugs in Covid. Within a day or two of my email threat, the chief executive editor of all the Frontiers journals (they have many in their collection), Dr. Frederick Fenter, informed me and Dr. Malone that he had received a complaint about our paper (I'm sure he had!) and as a result, he had supposedly assigned an anonymous third-party peer reviewer to appraise our data. Contrary to all of our prior esteemed peer reviewers, that anonymous arbiter determined that *our conclusions were not supported by our data* and they were going to retract the paper.

We were not provided with a written peer review, just a verbal notice of retraction. Together, in our careers, we FLCCC founders had passed peer review with more than 1,500 papers; this action was unprecedented. The single retraction suffered until this point among all of us was years ago when a professional rival of Paul's complained to a journal that Paul had "self-plagiarized" himself in one of his papers. Despite Paul's protest to the editor, his original paper was retracted. Paul then wrote a brilliant follow-up paper exploring the ridiculous concept of "self-plagiarizing." He titled it "Self-plagiarism: The perspective of a convicted plagiarist."[1] It's a hoot to read.

It's impossible to explain how insane this retraction was. Historically, the only reason for a paper to be retracted is when evidence of fraud or scientific plagiarism is uncovered. In our case, after passing peer review by four experts, suddenly some shadowy, anonymous fifth reviewer disagrees and says to retract the paper. And the editor listens to that reviewer alone. What?

Of critical importance is the fact that this reviewer did not detail anything specific to revise; he or she simply said that they "felt" the data presented did not support our conclusions and that it should be retracted and that further, *no opportunity to revise the paper should be offered*. The lack of an offer to revise the paper was, to me, definitive proof of their malevolence. Standard practice in scientific manuscript submissions is that when a peer reviewer has a problem with a paper's analysis or conclusion, they instead make a suggestion to revise it by weakening or limiting the conclusions or revising the method of analysis.

Sensing tragedy, I made a lame, pathetic plea by revising the paper to a more muted conclusion, hoping to satisfy Fenter and his anonymous reviewer (Hi, Bill!). I was half naive, half desperate, convinced that even a softened paper would be massive in impact. And as megalomaniacal as this sounds, I truly believed that this paper could *save the world*—or at the very least, a good chunk of it.

I still remember the stunned silence on the Zoom call with the FLCCC when I informed them of Fenter's final decision to retract.

Imagine being in a state where your knowledge or expertise could save millions of lives. I knew this paper had to be published or people would die. Lots of them. *It was all on me; literally, the weight of the world.* I swear I was as unsettled as you are at this thinking and its implications.

After our paper got retracted, Frontiers then put a hold on *all* the papers on repurposed drugs that were under review in Robert's special issue. Robert and his coeditors had a tense meeting with the journal editors as well as an outside expert who was brought in to review the submissions. During these discussions, it became clear to Robert and his colleagues that the whole process was fraudulent and that the "expert" was not making any scientific or ethical sense. Outraged, they resigned en masse, a scandalous move that was covered in a professional magazine for scientists aptly named *The Scientist*.[2] The special issue was dead.

Robert had extensive experience in designing, conducting, and publishing clinical trials as well as preparing pandemic responses with government and military funding. He was a veteran not just of science but of governmental science specifically. Robert became a counselor and mentor to me, and we would discuss how the rules of engagement had patently changed and debate our best tactical approaches. One day, I was struggling under the weight of it all. "I'm getting killed out here, man," I told him. "I could use some public support from a guy of your stature and credibility." Somberly, he replied, "If they can't see you, they can't shoot you."

His words were both chilling and accurate.

And here's the thing; this was before he was "ROBERT MALONE," the most public figure in our movement and the one with the biggest, brightest target on his back. At that time, he was a behind-the-scenes guy, trying to prod science in the right direction. It was his idea to propose the special issue on using repurposed drugs to treat Covid to *Frontiers in Pharmacology*, probably not something he would have pitched if he'd been aware that Frontiers was accepting funding from BMGF.

After our paper's surprise rejection, we licked our wounds and quickly submitted to another journal, the *American Journal of Therapeutics*. The editor of that journal, Dr. Peter Manu, had in fact reached out to me months prior and invited us to submit to him, but unfortunately, I declined, hoping to land a "higher impact" journal (I know!), and also because I wanted to get the paper into Robert's special issue, which I knew would generate considerable buzz. I actually knew Peter from my childhood—he had been friends with my parents—so there was a foundational layer of trust there, but my wounds were still fresh from the beating I'd taken at Frontiers, and I needed to know that Peter wouldn't pull the same nonsense.

Peter acknowledged the absolute unfairness of what we endured in the face of "an advancement in science," as he characterized it. It was simply the Semmelweis reflex; a term for the almost involuntary tendency to reject new evidence or knowledge because it contradicts established norms or beliefs (recall how Del Bigtree pointed out scientists originally bristled against hand-washing in the foreword of this book?). At least that was what he thought was occurring. His compassion and sympathy were immensely reassuring.

He asked that I submit Frontier's peer-review comments to him, including the history of our revisions. After careful review, he accepted our paper for publication.

Shockingly, it quickly became one of the most popular papers in recent history using the Altmetric Attention Score, an algorithm used to measure the relative reach of research articles. The AAS was only invented in 2011 so I don't want to overstate its importance, but if research were a popularity contest, we were killing it out there. Ours became *the eleventh most popular paper out of the last 23 million publications*; today it hovers around number fifty-eight.

The original actions taken against our ivermectin paper by Frontiers were not isolated. Within weeks of our retraction, Tess Lawrie's meta-analysis passed a full peer review at *The Lancet Respiratory Medicine* journal. Despite this, again, the editors refused to publish her paper, explaining their decision with this letter:

> Unfortunately, after some lengthy discussions with the editorial team, we do not feel that we can pursue the paper at *The Lancet Respiratory Medicine*. It was felt that there is just not enough evidence at the moment on ivermectin to be confident in a study such as this at this time, and we would encourage waiting until several more studies are published to help improve confidence in the paper. We don't doubt that this is an important paper, and would likely be widely picked up, and as such, we want to make sure that it includes as high-quality evidence as possible to ensure we spread a message that is strongly supported by the evidence. Therefore, on this occasion, we have decided not to publish your manuscript, but would perhaps consider an updated paper that includes more published evidence later down the line.

Not enough evidence . . . several more studies . . . decided not to publish. . . . It was Groundhog Day on steroids.

Tess followed our lead and submitted to the *American Journal of Therapeutics*. Her review paper[3] was accepted and became even more widely read than ours. As of today, it is the eighth most popular scientific publication of the past decade, out of 23 million. And yet there has still been no major media mention, even though this was occurring *before* all the counterfeit trials had been published. The signal of ivermectin's efficacy was profound and consistent, and at the time there was zero data to suggest otherwise.

Probably the most astonishing retraction was the study of Mexico City's bold early treatment program. In December of 2020, the city was approaching a state of collapse, with overwhelmed hospitals and medical supply shortages. The Mexican Social Security Institute (IMSS), one of the country's governmental public health agencies, designed a program focused on early treatment, widely distributing kits containing ivermectin in the midst of this unfolding humanitarian catastrophe.

The program was announced on December 29, 2020, weeks after the FLCCC's preprint review paper was posted, however it is clear they had been planning this well beforehand. The program included the deployment of 250 mobile testing units to the hardest hit neighborhoods of the city. Anyone who came for testing was provided access to a physician who could prescribe them a treatment kit. Further, they also initiated a telephone follow-up monitoring program for patients testing positive for Covid.

This is precisely what a public health response should look like in a crisis (and is also exactly what Paul Marik pleaded with the NIH to do eight days later to no avail. Even if the evidence had been "uncertain" for ivermectin (which it wasn't), such a program was justified on a risk/benefit analysis given the state of Mexico City's health system and the rising deaths being reported. They had to do *something*, damn it, so they did, unlike in the US, where, until pathetic Paxlovid came on the scene in mid-2022, health agencies and hospital systems were telling Americans to just wait at home until their lips turned blue. I wish I were making this up.

The Mexico City public health officials' post-program analysis found massive reductions in hospitalization rates among ivermectin-treated patients. The analysis was authored by some of the most senior public health officials in Mexico City and posted on a preprint server in May 2021. The only other ingredients in those kits were paracetamol (the same thing as acetaminophen or Tylenol) and aspirin.

To be clear, their program was not a clinical research trial; it was a public health program. It was not done to find out if ivermectin worked; it was done to try to save lives. It was not intended or structured as a research study. They sought input from numerous stakeholders, including *frontline clinicians with experience using ivermectin against Covid*. The public health officials trusted the collective wisdom and clinical expertise of this group and thus launched their program based on their communal assessment that ivermectin would be effective.

In Mexico City, they delivered 83,000 kits in the first month, which was the month that the study analyzed. It makes sense, no? If you launch a novel public health intervention program, it seems judicious to perform an analysis as quickly as you can to determine if it is working. The Mexico City program study was performed by the head of the Digital Agency for Public Innovation (DAPI), the Mexican Social Security Institute and the Mexico City Ministry of Health.

Performing an analysis of a program's impacts after the fact is a legitimate and long-practiced research method known as a retrospective observational controlled study. Research ethics allow for studies that review health data in hindsight to determine the associations of certain interventions with patient outcomes. You do not need a patient's informed consent because these studies are only observational, and because patient data is de-identified in the analyses. I have done many of these studies in my career. As long as you can demonstrate to the research oversight board that you have put into place safeguards to keep patient data secure, a waiver of informed consent is granted.

Deploying medical intervention programs based on supportive studies is how we improve the care we deliver to patients. Studying the effects of any program afterward is also a critical aspect of quality improvement efforts because you can identify aspects of the program that worked versus those that didn't. For instance, in the Mexico City study, they found that patients who participated in the telephone monitoring program had even fewer hospitalizations than the patients who did not. The only way to know if that program had been effective was by evaluating the data in hindsight; in doing so, they were able to expand it (versus scrap it). See? Quality improvement. It's a no-brainer.

Lead-authored by Jose Marino, the study's conclusion determined that "a significant reduction in hospitalization was found among patients who received the ivermectin-based medical kit; the range of the effect is 52 to 76 percent depending on model specification." In case you skimmed over that, I'll repeat: They found *a minimum of 52 percent and maximum of a 76 percent reduction in hospitalization* amongst 77,381 patients. Further, the latter estimate is likely the more accurate one given it was based on the most robust modeling analysis.

Once again, this finding absolutely needed to be blasted out to all the cities of the world. It should have been on the cover of the *New England Journal of Medicine* and the *New York Times*. Rachel flipping Maddow should have devoted an entire episode of her tiresome, predictable show to it. Of course, none of that happened. Instead, the paper sat on a preprint server for months. Compare the fate of this significant paper with those of the single studies of remdesivir, molnupiravir, and Paxlovid, whose results were immediately made public across the world by press release, before the public could see their trial data.

It was abhorrent, to say the very least.

We waited with the hopeful anticipation of a kid on Christmas Eve for the Mexico City paper to appear in a peer-reviewed medical journal, but it never did. Had that happened, even though "the zone" of media censorship was becoming increasingly powerful and consolidated, I believe it still would have made serious waves in at least some parts of the world.

Despite the widespread dismissal and ignoring of preprint servers, that paper was becoming a problem. Many of us were citing it tirelessly, me as much as anyone as I loved what that program accomplished. Unsurprisingly, after almost a year on the preprint server, someone (I suspect from Pharma/BMGF) decided to make a move, and they got the editor to retract the paper. *Off of a preprint server.* From the preprint server editors on their decision to retract:

> Our grounds for this decision are several:
>
> 1. The paper is spreading misinformation, promoting an unproved medical treatment in the midst of a global pandemic.
> 2. The paper is part of, and justification for, a government programme that unethically dispenses (or did dispense) unproven [there it is again!] medication apparently without proper consent or appropriate ethical protections, according to the standards of human subjects research.
> 3. The paper is medical research—purporting to study the effects of a medication on a disease outcome—and is not properly within the subject scope of SocArXiv [an online server for social sciences research].
> 4. The authors did not properly disclose their conflicts of interest.

Yes, public health officials whose conflict of interest was . . . public health? Clown World had appointed a new ringleader.

In a follow-up post[4] on the preprint server, the director of SocArXiv comically wrote, "We are taking this unprecedented action because *this particular bad paper* [emphasis mine; grammatical error theirs] appears to be more important, and therefore potentially more harmful, than other flawed work. In administering SocArXiv, we generally err on the side of inclusivity, and do not provide peer review or substantive vetting of the papers we host. Taking such an approach suits us philosophically, and also practically, since we don't have staff to review every paper fully. But this approach comes with the responsibility to respond when something truly harmful gets through. In light of demonstrable harms like those associated with this paper, and in response to a community groundswell beseeching us to act, we are withdrawing this paper."

This particular bad paper? Clearly the words of a deeply intelligent and well-informed individual.

The paper's lead author Jose Merino lashed back on Twitter.

José Merino @PPmerino · Feb 5, 2022   ...
Replying to @saiphcita @RicardoTBasurto and @familyunequal
You mean the open, public and anonymized data generated independently from and previous to the evaluation?

You mean the affiliations disclosed at the very beginning next to our names?

You mean the results that **you** could replicate for a whole year if **you** wanted to disprove them?

○ 89   ↺ 37   ♡ 32   ılıl   ↥   ⚠ Tip

Later, Merino and his team posted a more detailed and powerful defense of both the ethics of the program and their study, as well as how they carried out their analysis.

It didn't matter. "They" disappeared the paper from the preprint server. Cue the deafening media chorus of "all the ivermectin papers are fraudulent." The *Washington Post* gleefully ran a story headlined, "Mexico City gave ivermectin to thousands of Covid patients. Officials face an ethics backlash."

Given the incredible data showing the success of the program's first month, Mexico City continued to distribute the kit for over a year. *Because they knew it worked.* Unfortunately, even after they had seen firsthand how powerfully effective the ivermectin program had been, after the first two of the Big Six counterfeit trials were published, the program was canceled. On January 4, 2022, the IMSS announced they would no longer distribute ivermectin.

Naturally, the Associated Press was delighted to report this fact, its headline boasting, "Ivermectin is no longer in Covid-19 treatment kits in Mexico." The AP article contained this rationale for the program's halt: "In the statement sent to the AP, the IMSS said that ivermectin was dropped from the packets because of information contained in studies released this month by the World Health Organization and The National Institute for Health and Care Excellence in the United Kingdom."

It nearly made me cry, but I wasn't surprised. The program had to end. There was no way Pharma and BMGF would allow one of the world's largest cities to continue to have a sweeping early treatment program centered around ivermectin. Armed with the flawed and fraudulent "rigorous" trials emerging in the high-impact journals, they managed to somehow convince the presumably more politically powerful health officials to end the program. Sadly, this was modern medical science at work.

The next paper to get retracted was published by me and the FLCCC. It was a paper supporting our earlier MATH+ hospital protocol[5] that had been published in December of 2020 in the *Journal of Intensive Care Medicine*. Interestingly, ivermectin was not mentioned in that paper at all because at the time it was not part of our MATH+ protocol. Yet that paper was suddenly retracted a year later. Why?

Simple. Paul was the director of the main ICU at Sentara Norfolk General Hospital. He was using MATH+ and achieving mortality rates 50 percent lower than those of his colleagues who were relying on the ill-advised but widely accepted protocol of remdesivir and a pathetic 6 mg of dexamethasone. Paul was inadvertently calling attention to their—and the rest of the country's hospital systems'—horror show, which was a direct result of the ineffective NIH protocol. Desperate to destroy any evidence that the nation's pathetic protocol was inferior, they tried to silence him with something known as a "sham peer review." Despite sounding suspiciously like *peer review,* which is the scientific or academic evaluation of a fellow professional's research, a *sham peer review* is a malicious, barely legal abuse of the employee review process used to attack or depose a physician for personal or other non-medical reasons.

They basically ambushed Paul.

John Brush, Senior Medical Director of Research at Sentara, wrote to the Journal of Intensive Care Medicine and complained about how we presented our mortality data for MATH+ in the paper. It's worth noting that Sentara never claimed that we presented *false* data, but rather they objected to the denominator we used to calculate the mortality rate. The fact that it was the only denominator we had data for didn't matter. Nor was the method in which we calculated the mortality data different from several other papers published early in the pandemic. Alex Marinos investigated the journal's and Sentara's claims against us on his Substack,[6] and couldn't find that we did anything wrong. As well as highlighting countless nefarious deeds associated with the retraction, he pointed out that even if the paper were rewritten with the numbers the hospital preferred, it *still* would have shown a remarkable reduction in mortality. Marinos brilliantly asked, "*Why is the hospital asking for a paper to be retracted that demonstrates—by its own numbers—that it had substantially lower mortality than other comparable hospitals? Isn't that the sort of thing one brags about?*"

Apparently not, because the journal had already retracted the paper. Again.

The FLCCC has received many awards for our work in the pandemic. All of them are meaningful, but one stands out to me as a personal career highlight: That recall of that paper appeared on *The Scientist* magazine's "Top Retractions of 2021."[7]

CHAPTER TWENTY-EIGHT

The Journal Rejections of Positive Ivermectin Studies

The high-impact journal Editorial Mafia have another power equal to that of willingly publishing fraudulent trials, which is their ability to refuse to peer review any submissions of study manuscripts with inconvenient results for Pharma. This includes dozens of, in some cases, high quality, positive ivermectin trials.

The most damaging rejection has to be Tess Lawrie's systematic review and meta-analysis submitted to the Cochrane Library in January of 2021, within a month of my second senate testimony.

First, know that Tess and her colleagues Andy Bryant and Therese Dowsell are world experts in this type of study, with approximately 120 published Cochrane Reviews between them. Had Cochrane accepted her proposal and published her review, it would have neutered the ability of Andy Hill, Andrew Owen, and the WHO to issue their corrupt recommendations against ivermectin. Once a medicine has a Cochrane Library review supporting its use, it becomes established as the standard of care. A positive meta-analysis published there would have resulted in physicians the world over instantly adopting its use in their practices, saving hundreds of thousands of lives.

Initially, Tess proposed to Cochrane that her team do a rapid review of ivermectin. This is essentially an expedited, simplified review reserved for emergencies as it can be conducted more quickly than a full, systematic review. Cochrane initially agreed before suddenly backpedaling—again, likely due to Big Pharma/BMGF pressure. Suddenly, the Cochrane editors insisted that only a full review would be considered. Tess agreed and submitted a full review protocol as her team, in anticipation, had actually already completed the work.

The corrupted Cochrane Library was in a bind, so they pulled out the next weapon in their vast arsenal: "conflicts of interest" accusations, citing her video plea to Boris Johnson requesting he review her study results. Make no mistake: this *in no conceivable way* represents a conflict of interest. Tess does not make money from ivermectin. In fact, an expert of her level is morally and professionally obligated to share such findings to the public, especially during a global pandemic mired in a wicked winter surge.

To preserve the opportunity to publish such critical data in the vaunted Cochrane Library, Tess even offered to step down as an author to assuage their concerns of alleged conflicts of interest.

Her request fell on deaf (dumb?) ears. The editors simply told her to go publish in another journal. Cochrane was off the table.

The same story was playing out around the world. Dr. Eli Schwartz is a world-renowned professor of tropical diseases at one of the top universities in Israel. His high-quality RCT effectively "proved" the anti-viral properties of ivermectin against SARS-CoV2 when he found that both viral cultures and PCR tests cleared faster in those treated with ivermectin. I wrote to him many months ago asking when his landmark paper would be published; this was his reply:

Hi Pierre, the sequence of submissions were: NEJM, Lancet- eclinical medicine, and Clinical Infectious Disease. The rejections came within a few hours. At that time I did not submit it to medRxiv [the preprint server] to avoid rejection [by any future journals] based on "already published information." Now, we submitted it to Clinical Microbiology and Infection and in parallel to medRxiv.

It would have been a world-changing study. Although it was finally published recently in the *Journal of Infectious Diseases*, I had to look it up to find out. Its publication didn't even make a ripple in the media. Ironically, our old friend Gideon Meyerowitz-Katz has viciously attacked Schwartz's study for changing the protocol while the study was ongoing. Unlike the Big Six protocol manipulations, Schwartz did so *with a documented request and full approval of the research oversight board, and before any of the data were unblinded.* Note to Gideon: That's what real scientists do, you slimy bastard.

Things were no better in Argentina. Retired Professor and now friend Hector Carvallo submitted a paper to *JAMA* in early Covid showing massive impacts of an early treatment protocol he dubbed IDEA which centered around the use of Ivermectin, Dexamethasone, Aspirin, and Enoxaparin (a blood thinner), all of which have since been validated.

Hector received an immediate reply from the journal's editor-in-chief with a manuscript number and a promise of review. This is traditionally a highly positive sign as rejections, when they occur, typically result in a swift auto-reply letter.

His optimism was quickly crushed when he received the dreaded "auto-reply" not days or weeks but *hours* later. The reply read, in part:

> More than half of the approximately 7000 manuscripts submitted to us annually are rejected after such in-house review, and less than 9% of manuscripts are eventually accepted for publication in JAMA. Based on our evaluation, I regret to inform you that we will not pursue the manuscript you have submitted for publication.

This, in case I need to point it out, was highly irregular. I have never been told by a journal they would consider my manuscript for review and then been rejected later the same day. Typically, if the journal is interested in your work, peer reviewers are assigned; although this is no guarantee your paper will get published, it puts publication at least within the realm of possibility. It's virtually unheard of for a journal to open the door to publication and then immediately slam it shut.

Hector ended up publishing his impactful study[1] in the *Journal of Biomedical Research and Clinical Evaluation*, a publication that I had never heard of. Imagine winning the Nobel Prize and the only interview request you get is from some guy in his parents' basement launching his new YouTube channel.

Another casualty of this brutal, bloody war was the work of Professor Waheed Shouman of Zagazig University in Egypt. Shouman conducted a randomized controlled Covid trial[2] in which he found massive reductions in the incidence of infection among those taking ivermectin. It was a high-quality study from a reputable university, and it was rejected three times.

After multiple edit suggestions, it was rejected again, finally "after a long delay without revision."

That last part—*a long delay without revision*—was equally unprecedented before Covid. It's worth noting that it is against scientific journal policy for a study author to submit to two journals at the same time; you wait until you are rejected from one before submitting to the next. The journals know this. By "sitting on a paper" without rejecting it and without sending it for peer review, they essentially prevent the possibility of the dissemination of results. With a quick rejection, you can just lick your wounds and try with another journal. I cannot tell you how damaging this practice is to researchers . . . not to mention, to patients.

Shouman wrote to me a few months later, thanking me for my persistent defense of "a drug that won't sell to fill our pockets" (his words). He included a link to his published paper, which although historic, appeared in another publication—the *Journal of Clinical and Diagnostic Research*— that no one had likely ever heard of, and no media would cover.

Thousands of miles away in Nigeria, the same story was unfolding. Professor Olufemi "Femi" Babalola of Novena University had conducted a double blind RCT showing numerous statistically significant reductions in important endpoints among ivermectin treated patients. He sent it to the *Bulletin of the WHO*, the organization's monthly health journal, who replied with confirmation of consideration. His submission was rejected three days later—only Femi wasn't notified.

Femi was the first author and the submission author. When the WHO rejected it, they sent the rejection to *all of the other authors* . . . except Femi. He literally did not know it had been rejected, but his coauthors assumed that he did. This further delayed their resubmission to another journal. Honest error or low-down, carefully planned, dastardly little move? I'll let you decide.

As one of the principal investigators of a registered RCT of ivermectin, Femi was in contact with Andy Hill's team at the time. In a message to the principal investigators at large, Andy informed them that a senior editor with the *Lancet* journal *eClinicalMedicine* was interested in receiving new manuscripts for ivermectin RCTs. Andy acknowledged that getting these trials published had been a challenge—and was hoping this new connection could turn the tide.

This was back when Andy was still on our proverbial team. He was in contact with the principal investigators of all the active registered RCTs of ivermectin around the world. Like us, he was shocked that none of us could get our papers published.

Femi wrote back, informing Andy that when his submission was rejected by the *Bulletin of WHO,* they had submitted to the *Quarterly Journal of Medicine (QJM)* published by Oxford University Press and had "reasonable feedback from the assessors."

Andy was sympathetic. He lamented the volume of ivermectin papers being rejected and congratulated Femi on his paper, expressing hope that the process would become easier over time. It was the last truthful message any of us ever received from Andy Hill.

Two months later, Femi submitted his paper to the prestigious *Indian Journal of Medical Research.* They responded by saying that his study was being considered for publication. Over a month later, during the height of the Delta wave in India, he received this pathetic, generic rejection:

> Dear Sir/Madam,
> We express our inability to consider your article for publication. Thank you for submitting to the IJMR.
>   Editor-in-Chief

The deflect-delay-deny maneuver was gaining terrifying, deadly steam with the Editorial Mafia.

This next example of rejection fraud is perhaps the most frightening. Recall that preprint servers have likely saved millions of lives by quickly disseminating trial data of repurposed drugs. The FLCCC built our protocols by selecting effective compounds based on their mechanisms of action: *in vitro* (meaning studied in a controlled environment such as a test tube or petri dish), *in vivo* (studied in or on a living organism such as a person, animal, or plant), and clinical data (informed using research data, often sourced from preprint servers). In the latter case, we doctors "peer-reviewed" the papers ourselves, relying on the data to use and recommend therapies such as povidone-iodine nasal washes and gargles, melatonin, and vitamin D.

Femi attempted to post one of his newer ivermectin studies to the most popular medical preprint server, medRxiv. Understand that preprint servers are . . . *preprint servers.* As in, yet-to-be-peer-reviewed or published. As long as the study involves original data collected by the investigators, it gets posted after a cursory review to rule out ridiculous claims as well as verify the academic or research backgrounds of the authors. If you pass the smell test, your trial gets posted. Period. Well, unless it's an ivermectin study, apparently. The reply Femi received from medRxiv was devastating in countless ways. After a personalized salutation and some routine pleasantries, it stated:

> We regret to inform you that your manuscript will not be posted. A small number of papers are deemed during screening *to be more appropriate for dissemination after peer review at a journal rather than as preprints.* Please be assured that this conclusion is not a judgment on the merits of the work described.

It doesn't take a detective to determine that the preprint server editors got the memo to block any reports of ivermectin efficacy from being published. *Turf them to the Editorial Mafia, where they will further deflect, delay, and deny.* In support of my "two clicks to Bill Gates" hypothesis, the medRxiv homepage boasts that it is "supported by Chan Zuckerberg Initiative;"[3] BMGF partnered with Chan Zuckerberg on the COVID-19 Therapeutics Accelerator, a multimillion-dollar fund to "address the coronavirus pandemic."[4] Ahem.

Finally, I'll share a heartbreaking exchange I had with one of Hungary's top Covid experts, Dr. Zsuzsanna Rago, whom I got to know when she invited me to give a lecture remotely for a symposium she had organized. She has impeccable credentials: researcher; primary care physician; IVERCOV project leader in the University of Debrecen, Hungary; responsible for an ivermectin clinical trial in Covid patients and for the development of the Hungarian ivermectin generic drug.

Zsuzsanna wrote:

> Dear Pierre,
>
> Thanks for all your energy you put in fighting for early treatment in Covid-19. I follow your work at FLCCC. Last video Tess Laurie (you and Paul Marik) made as a letter to Andrew Hill is just great and for me is heartbreaking. We also did everything here in Hungary to introduce ivm treatment in Covid but the mainstream "science" *coming from* the western world was enough for our professors, high ranked doctors and authority to refer to and it pushed us back to the shadow. I still keep healing our patients anyway.
>
> Although in numbers it is very low, Hungary leads the Covid death rate in Europe.

Yesterday our ivm meta-analysis conducted by 3 universities of Hungary by my lead was rejected at editor level from Infection journal. The editor did not even send it to reviewers . . . .

We analyzed ivm efficacy on SARS-CoV-2 viral clearance in mild-moderate Covid. The result is (of course) significant in favor for ivm, the evidence level turned very low because only 3 studies fitted in our PICO. But the result is positive. The deadline of our search was 1/2021, so a year ago. And we are still not able to publish it.

Could you please recommend a journal who would accept our ivm paper without any censorship?

Thanks a lot,
Zs

Zsuzsanna Rago
Hungary

Could I please recommend a journal who would accept her ivermectin paper without any censorship? I would have laughed if I wasn't curled into a fetal position, sobbing.

As the world came under attack by a novel and highly transmissible respiratory virus, the medical journals had a critical, life-saving responsibility to publish any and all studies of treatments that were even potentially effective in preventing or treating the disease. They not only abdicated their primary duty but were influenced to violate it, repeatedly and egregiously.

Millions have died as a result of their willful and concerted actions.

If that enrages you, with all due respect . . . good. This army will take as many furious new soldiers as we can get.

PART THREE

THE AFTERMATH

CHAPTER TWENTY-NINE

Back in the Real World

A building is only as strong as its foundation, and the counterfeit footing of ivermectin science was built carefully and strategically. As brick after brick of scientific fraud was put into place, civil wars over ivermectin broke out in almost every country on the planet. That was the war that the FLCCC would find itself fighting in the wake of my second senate testimony. Not just in the US, but all around the world.

My elation over the viral spread of that testimony lasted all of about forty-eight hours.

I had begun my sworn statement by introducing myself as "part of a group of the most highly published physicians in our specialty." I was referring to the FLCCC, but apparently viewers took it to mean the major medical center in Milwaukee where I worked at the time (which I am unable to name as part of my severance agreement so I will henceforth refer to as Milwaukee's Beast). As my video made its way around the world, the center was getting bombarded with calls. Within two days, I'd gotten a call from the chief medical officer. He explained that they would be drawing up a new contract for me, and that I was not to talk to any press until we could meet.

I knew what was coming; they were either going to gag me or fire me. I told the CMO that if there was anything in the new contract that would restrict me from speaking publicly, it would be an extremely short meeting because I would not be signing it.

On a conference call that night, we were joined by the ICU chief who had hired me. It was not a pleasant conversation. They felt as if I had been speaking on behalf of Milwaukee's Beast. I insisted that it was clear from my remarks that I was not a spokesperson for the pseudonymous center and that my opinions were undeniably my own, but they felt otherwise.

They emailed me the new contract, which contained at least eight different restrictions on my first amendment rights. I refused to sign it.

It was Covid career casualty number two for yours truly.

I guess they don't call me "Lucky Pierre" for nothing. By this point, the FLCCC was doing well enough with donations that I could actually take a salary. I would meet so many warriors on the battlefield during this time who weren't so fortunate; outspoken physicians who lost their livelihoods for refusing to be silenced and were forced to create new careers and find new sources of income. The injustice of it all is staggering, and my hat goes off to each and every one of them.

Finally, it seemed, the FLCCC was gaining traction. I was beside myself when, a day later, I received a request for an interview by the Associated Press, self-described as "the largest news gathering organization in the world." This was huge—the global media homerun we'd been waiting for.

The AP dispatched a former fashion reporter named Beatrice Dupuy to interview me. I spent twenty minutes detailing the countless data points which consistently showed massive benefits with ivermectin treatment. The interview was cordial, and Beatrice appeared genuinely interested in and intrigued by the information I presented.

Shortly afterward, the AP ran their piece.[1] This was the headline:

"No evidence ivermectin is a miracle drug against COVID-19."

The article itself isn't fit for a birdcage. Beatrice deliberately omitted all the data I provided and chose instead to share the story of an Arizona couple who'd ingested a fish tank cleaning additive (chloroquine phosphate), which is an ingredient in hydroxychloroquine.

"The woman became gravely ill and the man died," Beatrice wrote breathlessly (I imagined).

Don Henley said it best: *"It's interesting when people die; give us dirty laundry."*

At the bottom of Beatrice's piece was this interesting disclaimer:

"This is part of The Associated Press' ongoing effort to fact-check misinformation that is shared widely online, including work with Facebook to identify and reduce the circulation of false stories on the platform."

The FLCCC promptly filed an ethics complaint with the AP. Thanks to an errant "reply all" on their part, we were able to see an email thread between the CEO, ethics chief, and president discussing a plan to delay their response so they could "buy some time" to figure out what to do. It's hilarious looking back at the naivete we possessed by *filing an ethics complaint against an erstwhile fashion reporter working for the Associated Press*. We actually believed that a moral code existed that we could rely on to force journalistic integrity.

Two weeks later we received a letter stating that the AP had investigated the complaint and found no ethical concerns with the piece. *As if they were actually going to side with us?* Talk about the fox guarding the henhouse. We had a lot to learn, but our optimism is amusing in hindsight.

Despite losing my job and the AP hit job on me occurring within days of my senate testimony, the FLCCC was kicking ass. Generous supporters like Jeff Hanson and Ronny Scerbar laid the foundation with large endowments, while smaller donations were increasing steadily. This financial support allowed the FLCCC team to finally be compensated for what until this point had been a passion project for many of us. Although the wages may have been more modest than many of us were accustomed to (I see you, Sean), nobody was complaining.

Still, achieving our mission of disseminating objective, sound, life-saving treatment guidance was a daily challenge. Between the chaotic, conflicting information on Covid therapeutics dropping daily and the slowly building censorship of effective, repurposed drugs, it was a 24/7 flurry of frenzied activity.

Sean Burke, our whip-cracking executive director at the time, had us meeting on a Zoom call *every single day* for months. The meetings would often last two or more hours as we juggled putting out PR fires (many around the first of the Big Six frauds), deflecting personal and professional attacks, and fielding and creating opportunities to spread our message.

I was on the phone for hours a day with my own team as well as a rapidly growing network of incredible humans around the world. Many were introduced to me by referral, some I found through my research, and others reached out to me first. We were pooling our knowledge and experiences and building a true hive mind. It was exhilarating and exhausting.

When I wasn't on the phone, I was at my desk reading, researching, and writing white papers distributed via press release every time a US or foreign health agency would issue their latest absurd policy recommendation. At the time, c19early.org had identified numerous treatments that had shown efficacy against Covid including ivermectin and hydroxychloroquine as well as povidone/iodine nasal sprays, melatonin, and aspirin, all of which were being ignored. In my spare time (sorry, family), I was keeping my eye on the vaccines, which were already proving to be a colossal trainwreck.

Pretty much every aspect of my life was neglected or gutted. My wife, my daughters, my health, my hobbies, my diet, and my friendships all took a backseat to Covid. I was eating too much, drinking too much, and not exercising at all. Unsurprisingly, I hit the heaviest weight of my life, smashing previous records by an inconceivable margin. It's equal parts shocking and agonizing when I look back at how intensely my personal life suffered during this time, but in those moments, I didn't feel as if I had a choice. I had a responsibility—lots of them in fact—and as one might say is par for Pierre's course, I didn't balance those responsibilities very well. I'm not making excuses; I'm simply pointing out that "work-life balance" is difficult (impossible?) to achieve or even prioritize when you're in a world war. People were dying and I couldn't and wouldn't walk off the battlefield.

CHAPTER THIRTY

A New Chapter for
Me and the FLCCC

In the six months after my December 2020 senate testimony, Sean and his team were unstoppable. We added to our growing ranks Christensen Public Relations, a talented troop of media all-stars who work with us to this day. I became close friends with the owner and founder, Steve Christensen, whose help and guidance have been lifesaving not only to me personally, but to Paul and the entire FLCCC. I've learned volumes from Steve and his carefully assembled team of some of the smartest and most committed policy advocates I know.

The next transformative chapter in the FLCCC's story opens with a force of nature named Kristina Morros. Kristina is an American nurse anesthetist who was living and working in the Netherlands and whose passionate advocacy caught my eye on social media. Multi-talented, deeply read on Covid, and not afraid to speak out, Kristina was an obvious choice to come work with us. Her roles in the FLCCC have been myriad, from moderating the Q&As on our weekly webinar to editing and polishing our published protocols and creating infographics on numerous aspects of our therapeutics. I also hired Lori Halverson, the best administrative assistant I have ever had, to help field emails and organize the database of physicians, nurses, and laypeople joining the FLCCC alliance.

The pace was relentless—and hasn't really let up in the almost two and half years since my testimony. At one point, Chris Martenson suggested we start doing a weekly webinar, which we continue today, sometimes hitting over 10,000 live viewers from various channels. We recently celebrated our 100th episode.

Imagine having a family and a full-time job and then piling on top of it a nonstop circuit of journalist interviews, podcast appearances, and lectures around the world. By the end of 2021, Paul and I had given talks to medical groups and health organizations in Puerto Rico, Nigeria, Mongolia, Ukraine, India, Malaysia, Netherlands, France, Brazil, Hungary, Italy, South Africa, Zimbabwe, and the Philippines. We also presented to agencies like the National Institutes of Health and the Defense Threat Reduction Agency (DTRA). Then came testimony and/or lectures to government bodies like the Philippines House of Representatives, the Canadian Parliament Conservative Caucus, Sri Lanka's President and Health Ministers, Republican Governors of the US, the House Freedom Caucus, and the Third Century Group. We lectured to large organizations like United Nations Correspondents Association, the Young Presidents' Organization (YPO), the NAACP, the Indigenous Peoples of the Amazon charitable organizations, and The Awakening with Bishop E. W. Jackson. Throw in a couple of august institutions like Harvard University and Kings of London University in the UK, and then add an endless string of Covid conferences and summits including the International Covid Summits in Rome, Vienna, and Bath, Canada's Semmelweis Clinical Conference, the International Lyme and Associated Diseases Society, and the Truth over Fear Summit, and you may begin to feel my exhaustion.

I was on the road more often than I was home. It's no wonder I was physically and emotionally unrecognizable.

On top of everything else, my ever-expanding network reached out to me constantly when they or their children, spouses, parents, friends, neighbors, or mail carriers contracted Covid. I was tele-diagnosing and calling pharmacies around the country (regular old retail ones in the beginning, and independent compounders later, as the war heated up) at all hours of the day and night. As a physician in a pandemic when so few were treating early, I could and would never say no to anyone in need.

And there were scores of people in need. Of the hundreds of patients I treated during this time, only two ever went to the hospital. One was a distant cousin who contacted me on day nine, already struggling to breathe. Despite three doses of ivermectin in conjunction with corticosteroids, HCQ and povidone/iodine spray, she went to the ER and was admitted for hypoxia. Although her access to ivermectin was promptly cut off in the hospital, still she came home five days later and was never ventilated. Only one patient died, despite very early treatment (I started him on meds on day two) but he was eighty-seven years old, on dialysis, with advanced heart failure. Nobody can save everyone, but I gave it all I had.

Until I started my private practice in February of 2022, I had treated every patient in the prior fifteen months pro-bono. And although now I only officially see patients within my fee-based practice, I have never charged anyone I know personally or even tangentially for care.

As a physician treating early with repurposed drugs during Covid, I was part of a tiny but hardworking minority. Fearlessly vocal Dr. Peter McCullough estimates that there were maybe 500 doctors in the country (including himself) practicing and promoting early treatment and whose efforts saved hundreds of thousands of lives. Dr. Ben Marble, a family medicine specialist in Florida, left the system early in Covid. Based on a spiritual vision he had one night, he started a tele-health practice called MyFreeDoctor where they would treat anyone at any time for free (he asked for a donation for those who could afford it). Dr. Marble's practice data shows they treated over 150,000 patients in 2021 alone. America's Frontline Doctors were also treating massive numbers. During August of 2021, at their peak, they reportedly treated 60,000 patients in that single month.

Despite my damn-near perfect track record, there were plenty of bumps in the road. I once treated a friend of a lawyer I knew, who then asked me to treat her sister and her parents. Although her father recovered from Covid, he became ill some weeks later and ended up dying of a condition which some non-awake sister of this friend-of-an-acquaintance concluded was the result of the corticosteroids I treated him with (for a short time as he was hypoxic and breathless). She then got the executor of the estate to hire a law firm to harass me. They haven't sued me yet—but they might try. (Hey, a guy sued McDonalds for making him fat—and won. Only in Clown World.) No good deed goes unpunished, apparently.

From December to May of 2021, even though I was clinically treating Covid in large numbers of outpatients, I began to feel like a "fake doctor" for not being in the ICU. I wasn't seeing or treating the more severe and challenging phases of the disease, and I missed the sense of reward that you can only get in the ICU. I was in my home office eighteen hours a day either on the computer or the phone, and I longed to get back onto the battlefield.

The answer was as insane as it was simple: I'd get another ICU job. You know, while running the FLCCC and my global campaign to disseminate the evidence of efficacy of ivermectin, speaking out in public every time I was asked (which was constantly), and writing in my sleep. I reached out to staffing agencies and accepted a job as an independent contractor at Aspirus Wausau Hospital two hours away in Central Wisconsin. My schedule consisted of three twenty-seven-hour shifts (that's not a typo) one week, with a full week off the next. With my two-hour commute from home, on workdays I'd leave the house at 5 a.m. for my 7 a.m. shift, work until 10 a.m. the following day and be home by noon for an afternoon of FLCCC meetings, podcasts, interviews, writing and finally, blessedly, sleep. Then I'd wake up at 5 a.m. and do it all over again.

I know what you're thinking: *WTF was I thinking?* But I absolutely loved that job. It was rewarding and challenging and most importantly, I had total autonomy. I was using the full MATH+ protocol with ivermectin; I could treat anyone with anything. *I could be a doctor, for God's sake.* My partners in the ICU who hired me were familiar with my work and appreciated the therapy protocols I used and the aggressive way I treated. Some had taken my ultrasound courses in the past; others knew me from using my point-of-care ultrasound textbook; the mutual respect was palpable.

I made great money at that job—and I earned every penny of it.

Back at the FLCCC, Sean Burke had busted his ass for us around the clock for six solid months, but eventually he had to get back to his business and his nonprofit. He was instrumental in building the plane, and now we needed a new pilot to fly it. We interviewed a handful of candidates and one stood out like a beacon of light on a bleak horizon. Kelly Bumann was not only a follower of the FLCCC but had been "doing her own research" and had come to the same conclusions we all had, not only on early treatment but also on the toxicity and lethality of the vaccines. A former VP of Marketing for STARZ cable and satellite TV network, Kelly had extensive experience in managing project teams, which we desperately needed. In addition, she is meticulously organized, financially savvy, and is a tactical and strategic thinker as well as a skilled manager.

Kelly jumped into the captain's seat mid-flight and never looked back, hiring an exceptional crew to assist her. This was no easy feat, as often when the applicants she was interviewing learned that we were "that fringe, radical, anti-vax organization," the exchange abruptly ended. That turned out to be a built-in filter that ultimately attracted the best people, the one whose hearts, minds and courage are what fuel the FLCCC; folks like Kate Vengrove, our Director of Development, whose dynamic fund-raising keeps us viable, and Zahra Sethna, a gifted wordsmith, public advocacy specialist, and branding expert who now serves as our Senior Director of

Communications. Zahra has become Kelly's right hand, and together the two power and direct nearly everything we do.

Finally, the FLCCC had assembled an elite, expertly trained army—which is exactly what you need when you're about to take on the formidable foe that is censorship.

CHAPTER THIRTY-ONE

Censored

Proponents of censorship will eagerly cite a long and noble list of its benefits: It protects children (and adults) from obscenity and pornography, preserves cultural values and traditions, safeguards populations against hate speech and discrimination, stymies slander and libel, protects national security, facilitates election integrity (ahem), and prevents the spread of "harmful misinformation."

Dictionary.com defines misinformation as false or inaccurate information, "regardless of whether there is intent to mislead." (*Dis*information, on the other hand, is always intentionally deceptive.) Since almost all of the Covid narratives were built on lies, the disinformationists gaslit the world into believing that anything counter to their story was what was untrue.

Covid was a war of information. On one side you had the powerful (and power-hungry) federal agencies, plutocratic globalists, and corrupt pharmaceutical companies on a mad quest for ever more money and control. On the other side, you had the Dissidents: a bunch of scrappy scientists and medical professionals with miniature megaphones in our hands and truth and morals on our side. When our anti-narrative science and data became too big, too powerful, and too irrefutable for the censors to manage, and our group of truthers too vast to discredit as the occasional fringe doctor or conspiracy theorist, the only thing left to do was silence us completely. Remove the platform and you remove the problem. Easy peasy.

To wit, we now know that the Department of Health and Human Services (HHS) paid media outlets $1 billion dollars to promote the safety and efficacy of the vaccines and combat "vaccine hesitancy."[1] Further, the

Biden administration has admitted to working directly with social media companies and encouraging them to be proactive in combating "misinformation." Most recently, it's been revealed that the same massive public relations firm, Weber Shandwick, worked simultaneously for Moderna, Pfizer, and the CDC, the latter of which paid up to $55.2 million for PR services, money which was paid in part using Covid emergency funds.[2]

We also know that the media formed a mega-monopoly called the (so-called) Trusted News Initiative (TNI), whose members included a few minor influencers like BBC, Facebook, Google/YouTube, Twitter, Microsoft, AFP, Reuters, European Broadcasting Union (EBU), *Financial Times,* the *Wall Street Journal,* the *Hindu,* CBC/Radio-Canada, First Draft, and Reuters Institute for the Study of Journalism. In 2020, partner members of the TNI agreed, in the words of Director-General Tim Davie, "to work together to ensure legitimate concerns about future vaccinations are heard whilst harmful disinformation myths are stopped in their tracks."[3]

Our little volunteer army was outmanned, outfunded, outgunned, and surrounded on all sides.

YouTube was one of the earliest and most aggressive censors. In May of 2020, the streaming platform published a "COVID-19 medical misinformation policy" which expressly banned any mention of ivermectin or HCQ in the context of Covid.[4] This was before any of the Big Six trials were published, mind you, not to mention before the many dozens of positive ivermectin trials had been made public. We were mere months into the pandemic, and *YouTube* was unilaterally deciding what information was and wasn't scientifically valid or worthy of discussion. Who was pulling the strings?

In July of the same year, Senator Ron Johnson had his YouTube account suspended for violating this ridiculous, Orwellian "misinformation" policy.

In a statement, Johnson stood firm, declaring, "YouTube's ongoing Covid censorship proves they have accumulated too much unaccountable power. Big Tech and mainstream media believe they are smarter than medical doctors who have devoted their lives to science and use their skills to save lives. They have decided there is only one medical viewpoint allowed and it is the viewpoint dictated by government agencies."[5]

This censoring of scientific opinion was so absurd, I still turn purple with rage just thinking about it. YouTube forbade any discussion of two safe, widely used medications while their efficacy in Covid was just beginning to be studied and debated. In the midst of a global pandemic with thousands dying each day, a godforsaken *user-populated home-movie platform* was deciding what you could and couldn't hear. "You" Tube, my ass. It was ThemTube all the way.

COVID-19 medical misinformation policy

If you're posting content

Don't post content on YouTube if it includes any of the following:

Treatment misinformation:

- Content that encourages the use of home remedies, prayer, or rituals in place of medical treatment such as consulting a doctor or going to the hospital
- Content that claims that there's a guaranteed cure for COVID-19
- Content that recommends use of Ivermectin or Hydroxychloroquine for the treatment of COVID-19
- Claims that Ivermectin or Hydroxychloroquine are effective treatments for COVID-19
- Other content that discourages people from consulting a medical professional or seeking medical advice

Prevention misinformation: Content that promotes prevention methods that contradict local health authorities or WHO.

- Claims that there is a guaranteed prevention method for COVID-19
  - Claims that any medication or vaccination is a guaranteed prevention method for COVID-19
- Content that recommends use of Ivermectin or Hydroxychloroquine for the prevention of COVID-19

Facebook, Instagram, Twitter, and LinkedIn followed suit. You literally had the government and Pharma pressuring all the mainstream and social media giants to remove mentions or discussions of any and all early treatments—including (and I am not making this up), *prayer.*[6] The latter, according to YouTube's "Covid-19 medical misinformation policy," could be construed as "treatment misinformation" and result in the removal of your content.

Some platforms were more punitive than others, and all were claiming to be adhering to the guidance set forth by the WHO. This required two wild and erroneous assumptions: One, that the WHO was infallible; and two, that it hadn't been completely captured by Big Pharma and BMGF over the past few decades. There's a must-watch documentary called *Trust WHO* that exposes the agency's hidden history of dirty practices and will enlighten (and likely infuriate) anyone who's interested.[7]

Twitter was notorious for removing researchers, scientists, and experts who went against the government grain. (Brad Skistimas, the wildly talented singer/songwriter behind the band Five Times August and an outspoken Covid ally, dubbed it "getting sent to Twitmo." Sidenote: My coauthor Jenna and I would both like to have Brad's song "I Will Not Be Leaving Quietly" played at our respective funerals. Also, I suppose I should add that we are both in perfect health and have no desire to die anytime soon.) Notable Twitmo prisoners have included FLCCC analyst and ivermectin expert Juan Chamie, Robert Malone, Steve Kirsch, Alex Berenson, Peter McCullough, Representative Marjorie Taylor Greene, Naomi Wolf, Mary

Beth Pfeiffer, Daniel Horowitz, and Covid Crusher. Remember Covid Crusher? A bigwig in the biotech space who had to post pseudonymously, he was a beast early in the pandemic, constantly breaking new studies and sharing analyses on all aspects of Covid, including ivermectin. I learned a ton from that guy.

The recently (spring 2023) released "Twitter Files" shed eye-popping light on just how scandalous Covid censorship was. Author and award-winning journalist Matt Taibbi was one of a handful of reporters Elon Musk granted access to the platform's internal pandemic communications. Included in the files were weekly briefings from the Virality Project, a Stanford University research effort "aimed at understanding the disinformation dynamics specific to the COVID-19 crisis."[8] In reality, Taibbi's research revealed, it was a covert and sweeping censorship campaign involving billions of social media posts monitored by Stanford, federal intelligence and health agencies, and an army of (often state-funded) nonprofits. This wasn't just happening on Twitter, mind you, but across the entire swath of powerful social media platforms including Google/YouTube, Facebook/Instagram, Medium, TikTok and Pinterest.[9]

The magnitude of such a massive and coordinated attack on free speech and the dissemination of critical information during a global emergency simply cannot be overstated.

Among the most egregious offenses: *Any* "true content which might promote vaccine hesitancy" (including but not limited to a user posting about his or her vaccine side-effects or those of a loved one, a celebrity death after vaccine, natural immunity, campaigns against vaccine passports, "purported links between mRNA COVID-19 vaccines and cancer," or the closure of a school due to reports of post-vaccine illness) was considered "actionable content," meaning it could be censored or removed. In a testimony before the House Judiciary Committee, Taibbi accused the mainstream media of being "an arm of a state-sponsored thought-policing system," and of creating "a form of Digital McCarthyism."[10]

Talk about an invisible enemy.

Although the Dissidents were clearly losing in mainstream and social media circles in these early months, we were killing it in the independent news world—particularly podcasts. In the information war, those guys were beasts. In the US we had rockstars like Joe Rogan, Bret Weinstein, Chris Martenson, Russell Brand, Jimmy Dore, and Dr. Been forming a rebel news militia committed to giving truth a platform. Their international colleagues included John Campbell in the UK, Trish Wood in Canada, and New Zealand's Peter Williams, a veteran TV and radio broadcaster on par with

Walter Cronkite. I did two interviews with Peter and only learned months later that his forty-year media career ended one day after the second one.

In my travels across the country and around the world, people tell me all the time, "I first woke up after hearing you on Rogan or Weinstein." Those podcasts were huge, and I credit Bret and his wife Heather Heying for that. Both are authors and professors with PhDs in evolutionary biology; they are also lifelong liberals, outspoken egalitarians, and prominent intellectuals. In addition to hosting their popular DarkHorse Podcast, in 2021 they published *A Hunter-Gatherer's Guide to the 21st Century*.[11] It quickly became a *New York Times* bestseller.

Heather and Bret came across an article by *New York Times* bestselling author Michael Capuzzo titled, "The Drug That Cracked Covid,"[12] which featured first-hand accounts of ivermectin's lifesaving abilities. They had seen my testimony and began researching the therapeutic and were troubled that none of what was happening around ivermectin made scientific sense. We were following them, of course, because they were discussing ivermectin in the spring of 2021 in an objective, credible fashion. We later discovered Bret had over 500,000 Twitter followers and millions of podcast views. When Bret reached out to me through the FLCCC website we hit it off immediately, and he had me on his show on June 1, 2021. That single podcast got hundreds of thousands of views on YouTube alone before they took it down and Bret got demonetized. Bret knew how important the ivermectin science was, and since he was friends with Joe Rogan and Joe trusted him, three weeks later he and I had a nearly three-hour chat with Joe on his podcast, The Joe Rogan Experience.[13]

Those podcasts reached millions of people, despite the social media mafia's attempts to stymie the spread of information. While podcasts in particular were disappearing from YouTube, people were switching to censorship-free platforms like Rumble and Odysee and, in Joe Rogan's case, Spotify. Using their own secure, independent websites and these alternative platforms, independent podcasters like John Campbell, Mobeen Syed, Joseph Mercola, Greg Hunter, and countless others worked tirelessly to disseminate life-saving information to people around the world.

The independent publishing site Substack was, for me, the single greatest supplier and distributor of scientific information. Researchers could post their detailed analyses and cite their sources easily with a click. Steve Kirsch and Robert Malone had started writing detailed, striking Substacks containing a wealth of expertly researched data. Steve is a serial entrepreneur, philanthropist, and according to his Wikipedia bio, "a promoter of misinformation about COVID-19 vaccines,"[14] like the rest of us. He not only founded seven different tech companies but also is the co-inventor of the optical mouse. At the height of professional success, Steve developed a rare, deadly blood cancer with an estimated survival rate of five years. He assembled and funded a team of top scientists who identified a repurposed drug (Imbruvica) to treat it. He is still alive and slaying giants sixteen years later.

Steve went on to become a philanthropist, funding environmental, medical, and local causes to such a notable extent that Hillary Clinton presented him with a National Caring Award in 2003. (That honor was later scrubbed from his Wikipedia page, because these days if you don't like history, you just rewrite it.) In April 2020, he donated $1 million of his own money and raised millions more from donors to establish the Covid-19 Early Treatment Fund (CETF), which funded clinical trials of several repurposed drugs for Covid, most notably fluvoxamine. He did all this long before he was dubbed an "anti-vaxxer" (despite having been vaccinated). Months after the global vaccine launch, Steve was disturbed by alarming data that was emerging highlighting the toxicity and lethality of the vaccines, as well as the number of people in his social circle reporting injuries after vaccination. In response, he formed the Vaccine Safety Research Foundation (VSRF), whose mission is to advance Covid vaccine safety through scientific research, public education, and advocacy, and to support the vaccine injured.[15] Steve was on the battlefield from the very beginning, leading the charge against vaccine mandates and sharing valuable research through his Substack.

I followed Steve and Robert's lead and started exposing the fraud around both early treatment and the vaccines in my own Substack. I currently have 65,000 subscribers and my posts regularly get well over 50,000 reads. The FLCCC and many others also tightly secured our websites and hosting servers and watched them explode in popularity. We were getting millions of website views a month and had thousands watching our weekly webinars live from around the world.

Still, you had to be looking for us to find us—and even then, it could be a challenge. In an interview where we discussed ivermectin, journalist Matt Taibbi referred to me as "the Ghost of The Internet," because as soon as a researcher, podcaster, or journalist interviewed me, the video would be promptly taken down. In some cases, the other party would be de-platformed or demonetized, as happened to the brilliant medical educator Dr. Been, who (along with John Campbell, Bret Weinstein, Steve Kirsch, Paul Marik, and others) was a critical source of sound, accurate medical information in the pandemic.

The speed at which I disappeared has been astonishing at times. I recently gave a lecture at an IV vitamin C conference based on my pre-Covid research in sepsis. I did not even mention ivermectin or Covid, yet within an hour of the organizer posting my lecture on YouTube, it was taken down.

Later, the FLCCC would catalog the censorship actions against us. That list includes (but is not limited to) getting locked out of Twitter, YouTube, and Facebook (twice); being completely deplatformed on Medium, Linkedin, and Vimeo; and being banned from PR Web/PR Newswire and PayPal. The PayPal one really hurt as that was our fund-raising source. They literally tried to starve our organization to death.

One epic moment of censorship buffoonery was when, in another shockingly Orwellian move, the US Department of Homeland Security announced the creation of a Disinformation Governance Board to, "combat misinformation, mal-information, and disinformation that threatens the security of the homeland." They claimed that (our) contrarian Covid opinions had the potential to incite violence, which effectively made us domestic terrorists. Yes, possessing and sharing *scientific, medical knowledge was a threat to national safety.* The very idea of a Ministry of Truth in the US was a direct strike on our First Amendment rights, and fortunately, the criticism was swift and sweeping. The board's dramatic, Mary Poppins–singing director Nina Jankowicz lasted all of three weeks before resigning, and the group was dismantled less than four months after it was introduced.

At least in the case of the Disinformation Governance Board, they were broadcasting their wicked agenda—and we knew who was behind it.

Alas, as they say, prisons build better criminals.

Case in point: In April 2023, the *Epoch Times* ran a harrowing story that reads like a dystopian novel.[16] In it, they expose the insane inner workings of a group (Shots Heard) which is part of another group (The Public Good Projects[17]), that is home to a partnership (the Public Health Communications Collaborative), which counts the CDC Foundation as one of its member organizations. These groups' collective mission statements include the terms *online safety, digital cavalry, media monitoring,* and *social change interventions,* which hopefully tells you all you need to know. The goal of the Collaborative is to "decrease misinformation and increase vaccine demand worldwide," and they are using *social media influencers* (oh, how I hate that phrase) to do it. The doctors, nurses, and pharmacists in these groups actively encourage one another to report and harass "misinformationists," and openly boast/celebrate when their tattling gets the people or posts they dislike abused or shut down. I wish I were making this up—or that that was the worst of it.

Did you notice how the whole sorry lot funnels right into the CDC Foundation? That foundation is ostensibly an independent nonprofit organization established by the US Congress to support the CDC, making bribery via massive donations just-barely this side of legal. The foundation "provides funding and resources to the CDC to support a wide range of public health initiatives." (Like vaccines for everyone, for instance.) Their donors and sponsors—conveniently listed right on their website[18]—include, shockingly (not), the Bill & Melinda Gates Foundation, GAVI Alliance (hi, Bill!), Pfizer, AstraZeneca, Johnson & Johnson, Merck, Facebook, Google, Microsoft, and PayPal, among others.

In case you're missing the giant, flashing, neon point here: The deep-pocketed, pro-vaccine, anti-ivermectin players can and do "donate" massive amounts of money to a government-sponsored "nonprofit foundation" aimed at *improving public health.* If that statement doesn't make you want to move to Denmark or learn to grow medicinal herbs, I will consider this book a colossal failure and go back to kitesurfing full-time.

So yeah, the United States of Pharma is rallying its best Twitter and TikTok assets to destroy anyone like me, the FLCCC, or any of the heroes highlighted in these pages, and it's pretty clear they have no intention of letting up.

Unfortunately for them, neither do we.

CHAPTER THIRTY-TWO

Science Becomes Politically Polarized

Although the censors were relentless on the internet and on social media, the wider media landscape was a bit more forgiving—with a strange political twist. While the team that made up the FLCCC was almost uniformly liberal Democrats at the beginning of Covid, we had zero political agenda. Our mission was to develop treatment protocols for all aspects of the disease and to disseminate that guidance to the public. Yet for some reason, the only media interested in any of us or our cause came from the conservative side of the political spectrum. (Early in the war, I actually turned down an interview with Tucker Carlson out of fear of appearing too politically partisan.) Ironically, by being contrarian to the administration in charge of the captured and corrupted health agencies, the "right-wing media" was getting everything right. I'm not convinced that had Republicans been in charge, things would have turned out that much differently. Maybe a little . . . which would have meant a lot.

Meanwhile the left was publishing endless hit jobs on us, our message, and our guidance. The *New York Times, Washington Post, LA Times*, CNN, Rachel Maddow, and the entire flock of late-night talk show hosts seemed to be reading from the same script. Because our message deviated from that narrative, we were depicted not as scholarly and objective authorities on Covid science, but as doctors with a political or profit agenda. The systematic smear campaign was unprecedented. Political ideology had nothing whatsoever to do with my medical practice, yet it wasn't long before I was being called a "right-wing, fringe doctor."

At least there were outlets willing to give airtime to rebels like me and my team. The short list of supporters included Newsmax, One America Network, and Fox News hosts Tucker Carlson, Laura Ingraham, and my favorite, Maria Bartiromo. Maria in particular impressed me not only with the way she prepares exhaustively for interviews, but with her curiosity, accuracy, and investigative skills as well. She is a true professional.

As was Eric Bolling. I was a guest on Eric's daily Newsmax show, The Balance, more times than I can count, seemingly once a week for months. But initially, we were all uncomfortable accepting interviews from politically partisan outlets—especially the ones that leaned hard to the right. I had been indoctrinated to believe conservative media contained only distorted reality or outright lies; who wanted to be associated with that?

Over time, we had to face the reality that our responsibility was to the public, and if the only part of the public willing to listen was on the right side of the political spectrum, so be it. Both red and blue needed to be saved from the lies, but the blues largely refused to listen to anything from a non-establishment source. And we were clearly and vociferously anti-establishment.

That is also when I realized that the tables had turned. Left was now right, and right was left. And the new left had been duped on a colossal scale. The old "power to the people, stick it to the man" crowd suddenly believed the government was going to step in and save us all, while not realizing the government had been completely captured by industry. They were (and are) rooting for a corporate fascist state, while "right-wing nuts" like me became actively anti-government.

We were in an information war, with the disinformationists (them) fighting against the misinformationists (their name for us), the key difference between the two terms being that the latter is more of an "oops, we got that wrong," while the former engage in the deliberate spreading of lies. At the time, the disinformationists were winning, because they had the media (and "the science," meaning America's top infectious disease expert for seven presidential runs) on their side. Although the title they gave us is a blatant lie, I am proud to be part of the group fighting back, one I prefer to call the Dissidents.

Many of our followers have told us that we saved them; not just with ivermectin, but with sanity, common sense, and fellowship. We were (and are) highly credentialed and accomplished physicians who spoke plainly and sensibly and most of all, objectively. We were citing the data we had compiled using a "totality of the evidence" approach featuring numerous sources, and not-over relying on captured Big Pharma journals and agencies.

People have told us that they were terrified and confused until they found us; that the FLCCC feels like an island of sanity in a sea of madness. These things are gratifying and humbling to hear, but we aren't in this for the kudos. As career educators, we're all just doing what we have always done and always will do: asking questions, researching thoroughly, analyzing carefully, applying prudently and, most of all, sharing widely.

Each day, support for the FLCCC and our work and mission grows. Although this backing continues to come mostly from only one side of the political aisle, that side seems to be gaining some serious traction of its own, adding daily to its ranks people like me, who were raised squarely on the left but suddenly and unexpectedly, identify wholeheartedly with the right.

It could be because the right, on occasion, actually admits to being wrong. One of the most powerful examples in modern history of a major media personality calling out the fact that almost the entirety of mass media is beholden to only reporting narratives that support the interests of their Big Pharma sponsors was Tucker Carlson's monologue on April 21, 2023. The country's most popular cable news anchor was promptly fired, making that his last—and most memorable—show ever.

My friend Louisa Clary, the executive director of Steve Kirsch's Vaccine Safety Research Foundation, shared a meme with me once that made me laugh, the gist of which was this: "Who do you think believes more in penicillin, Republicans or Democrats?" It's funny because it's a ridiculous question, namely because *penicillin has nothing whatsoever to do with politics.*

Sort of like Covid.

CHAPTER THIRTY-THREE

The Horse Dewormer Campaign

Once the opposing army silenced all of the dissenting voices, the next step was to carefully craft and disseminate the approved messaging: *ivermectin doesn't work; all of the studies are too small, of poor quality or fraudulent; it can't reach effective concentrations; the only studies showing any benefit are in worms*. The propaganda campaign was relentless, running on every newspaper, TV show, radio station and social media platform simultaneously.

The problem (for them) was that people still wanted ivermectin. Prescriptions in the US were off the charts—literally. This graph shows how ivermectin had exploded in popularity by the middle of the brutal US Delta wave in mid-August 2021.

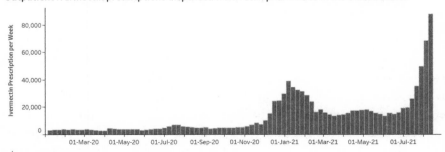

Outpatient ivermectin prescriptions dispensed from retail pharmacies in the United States

Data extracted from CDC Health Alert Network August 26, 2021, CDCHAN-00449
Elaboration: Juan J Chamie

Figure 14. (Source: CDC Health Alert Network)

Big Pharma and their big health agencies saw that US doctors across the land were increasingly and effectively treating Covid with a generic, repurposed drug. I believe it was this prescription data that unsettled the other side, because the timing was utterly predictable. They needed to bring in the big guns—a clever catch phrase or jingle or meme perhaps that would spread like wildfire and shut down the whole ivermectin business for good.

Cue Mr. Ed.

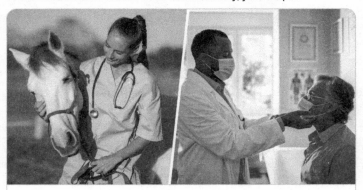

> **U.S. FDA** ✔
> @US_FDA
>
> You are not a horse. You are not a cow. Seriously, y'all. Stop it.
>
> fda.gov
> **Why You Should Not Use Ivermectin to Treat or Prevent COVID-19**
> Using the Drug ivermectin to treat COVID-19 can be dangerous and even lethal. The FDA has not approved the drug for that purpose.
>
> 6:57 AM · Aug 21, 2021
>
> **48.4K** Retweets   **22.3K** Quotes   **110.8K** Likes   **2,947** Bookmarks

"You're not a horse," debuted on August 21, 2021, with a tweet by the FDA, two months after I appeared on Joe Rogan's podcast and we talked about ivermectin at great length to Joe's (at the time) upwards of 11 million listeners. After the manipulation of the Pharma-funded trials, the horse dewormer nonsense was the war's second greatest offensive. It is my unsubstantiated belief that Weber Shandwick, the PR firm working simultaneously for Moderna, Pfizer and the CDC, had constructed the campaign months earlier and was just waiting for the best time to launch it.

Five days after the FDA's "tweet heard around the world," the CDC followed up with a memo issued to every state health department in the country, which then sent it to every licensed doctor in their state. "THIS IS AN OFFICIAL CDC HEALTH ADVISORY," it said in bold orange type across the top; the headline beneath that read, "Rapid increase in ivermectin prescriptions and reports of severe illness associated with the use of products containing ivermectin to prevent or treat Covid-19." Within twenty-four hours, every single licensed physician in the US had this memo in his or her inbox.

The problem was it was basically bullshit.

If you break the headline down grammatically, the first half is literally saying that "a rapid increase in ivermectin prescriptions [is] associated with the use of products containing ivermectin." Not exactly a work of linguistic brilliance, but accurate, nonetheless. The second half of the headline suggests that "[a rapid increase in] reports of severe illness [is] associated with the use of products containing ivermectin."

Investigative journalists Mary Beth Pfeiffer and Linda Bonvie decided to look into this claim. After a brief investigation, they penned a brilliant takedown of this fake media attack.[1] In it, they revealed that the entire "health advisory" was based on the single state of Mississippi's reporting on ivermectin-related calls to the state's poison control center. Do you want to know how many calls they had related to ivermectin in the entire state? Six. Of those six, the number of calls specific to the veterinary version of ivermectin was . . . four.

Four calls. Not overdoses, deaths, or emergency requests; just calls.

Again, this health advisory was sent to roughly a million physicians in the US, and the tweet, according to Pfeiffer and Bonvie (whom I must give props for using "Horse-Bleep" in their headline), reached as many as 23.7 million people in two days.

But liars are gonna lie, and they had a narrative to protect, so three days later, they trotted Fauci out on CNN.[2] Following Jake Tapper's hyperbolic opening that "poison control centers are reporting that their calls are spiking in places like Mississippi and Oklahoma because some Americans are trying to use an antiparasite horse drug called ivermectin to treat coronavirus," America's bobblehead couldn't wait to reply with brazen anti-ivermectin BS. "Don't do it," Dr. Do-Little commanded his minions solemnly. "There's no evidence whatsoever that that works, and it could potentially have toxicity." It was a perfect echo of the memo Merck posted on their website seven months earlier. To fully understand the scale of Fauci's nationally televised lie, as of a month earlier, there were sixty controlled trials showing ivermectin's consistent and robust efficacy.

Interestingly, that very same month, Fauci sat down for an interview with Dr. J. Stephen Morrison, senior vice president at the Center for Strategic and International Studies in Washington, DC, and director of the Global Health Policy Center. When Morrison asked Fauci to describe the optimal product profile for a Covid therapeutic, Fauci replied, "I want a pill that blocks a specific viral function. I want to give it once a day if possible. I want it to be low in toxicity. And I want it to have very minimal drug-drug interactions. So orally administered, single pill, given for seven to 10 days, little drug-drug interaction, and low toxicity; give me that and I'll be really happy."[3]

He literally described ivermectin.

Two short days after Fauci's CNN interview, three major medical societies (the American Medical Association, The American Pharmacists Association, and the American Society of Hospital Pharmacists) threw a rabbit punch. Instead of simply repeating the "warning" against ivermectin use like the CDC and FDA did, they instead put out a "call for an immediate end to prescribing, dispensing, or using ivermectin to prevent or treat Covid."

That shoutout was carried by every major news organization around the world, including our friends at the Associated Press. It wasn't just an unprecedented action (ivermectin had been FDA approved for years); it was a joke. *These agencies had no authority to issue such a call.*

At the risk of repeating myself, three major US medical societies were calling for an immediate end to the use of a medicine supported by a meta-analysis of sixty controlled trials showing it leads to major mortality reduction and other benefits. Now you know why I call our country the United States of Pharma.

For the next month or more, you couldn't turn on the TV, scroll through social media, or tuck in for a little late-night entertainment without seeing or hearing the words *horse dewormer*. It was all talking heads like Jimmy Kimmel, Seth Meyers, Stephen Colbert, and Rachel Maddow could talk about. And then, smack in the middle of this manure pile of a propaganda campaign, there was an insane plot twist:

Joe Rogan got Covid.

By the time he announced that he'd contracted the virus, Joe had already been treated with a combination of therapies, including ivermectin. The media lost their ever-loving collective minds. "Joe Rogan Says He Tested Positive for COVID-19, Takes Unproven Horse Dewormer," *The Hollywood*

Reporter bellowed. (Subhead: "Ivermectin has been falsely touted by some anti-vaxxers as an alternative to getting vaccinated.") Of course, no reporter bothered to explain that the uber-popular podcaster took the human version prescribed by his doctor. And when CNN decided to replay Rogan's self-recorded video of his experience, they added a creepy yellow tint to his face, making him look deathly ill.

The best was yet to come. *Rolling Stone* magazine published a one hundred percent made-up article (I am not making this up) with this headline: "Gunshot Victims Left Waiting as Horse Dewormer Overdoses Overwhelm Oklahoma Hospitals, Doctor Says." You know how they say that a lie can travel halfway around the world while the truth is still putting on its pants? By the time (two days later) *Rolling Stone* admitted the story wasn't true and printed their worthless retraction (confirmed by CNN even, another pandemic first), the original article had gone mega-viral. The icing on this ridiculous Clown World cake: In a tweet promoting the article, *Rolling Stone* included a picture showing the perfectly perky alleged gunshot victims waiting in line . . . wearing winter coats. It was late summer in Oklahoma at the time.

Rolling Stone ✔
@RollingStone

Gunshot victims left waiting as horse dewormer overdoses overwhelm Oklahoma hospitals, doctor says
rol.st/38CChjl

3:38 PM · Sep 3, 2021 · SocialFlow

6,407 Retweets **9,880** Quote Tweets **18.1K** Likes

Despite the absurdity of it all, the deadly horse dewormer narrative was here to stay.

In a convenient twist of Pharma fate, six weeks after the FDA posted the most famous tweet they will ever have solidifying ivermectin as a farm animal favorite, the media raced to launch their campaigns promoting Pfizer's Paxlovid and Merck's molnupiravir for the treatment of Covid.

Not only was the timing an especially fortunate coincidence, but in one of the greatest ironies of all time, it turns out molnupiravir is also used to treat *equine encephalitis*. In horses.

CHAPTER THIRTY-FOUR

The Covid Career Graveyard

Truth is the enemy of propaganda. And that is precisely why propagandists go after the truth-tellers. In Covid, they did this by branding anyone whose opinion differed from the United States of Pharma narrative a misinformationist, and then attacking their credibility with labels like "fringe," "right-wing," "fraud," or "quack."

Indeed, as Robert F. Kennedy Jr. so aptly summarized it, "The phrase 'medical misinformation' about COVID-19 seems to be a euphemism for any statement or scientific evidence that differs from the prevailing narrative of the vaccine and patented drug stakeholders."

In the article detailing common disinformation tactics, the harassment of scientists or experts who speak out with results or views that are inconvenient to industry is called a "blitz." Although blitzing is at its heart a censorship tactic, it often reveals itself in the form of propaganda. Can't shut someone up? Convince the world they're not worth listening to and you won't have to. Problem solved.

This is not a new tactic for Pharma. A CBS News expose[1] in 2009 detailed how Merck made "hit lists" of doctors who criticized their deadly, but for a time, blockbuster anti-inflammatory drug Vioxx. These lists were categorized with actions like "destroy," neutralize," or "discredit."

You: *A publicly held, global pharmaceutical company wouldn't do such a thing!*

Me: *Let me tell you about the time Merck was found guilty of creating a fake medical journal,[2] filling it with pro-Vioxx studies, and then distributing it to doctors across Australia.*

In the five years it took to get Vioxx removed from the market, the drug was estimated to have caused 100,000 heart attacks and 55,000 deaths. In senate testimony from 2004, Dr. David Graham, the Associate Director for Science and Medicine in the FDAs Office of Drug Safety, described the carnage as the equivalent of "two to four jumbo jetliners crashing every week for five years."[3]

It would be bad enough if I were merely saying that Pharma doesn't care about those side effects and deaths (they don't). But what I'm saying is far worse: They knew about them all along and did everything they could to suppress negative outcomes, falsify positive ones and generally, buy themselves as much time on the market as possible before an inevitable shutdown. When you're raking in millions of dollars a day, every minute counts. It's vile, reprehensible, and reason enough to hope there's a tenth circle of hell reserved especially for dirty Pharma scoundrels.

But in the fear-soaked pandemic frenzy, decades of pharmaceutical transgressions were forgotten. I guess if you were drowning and the only person offering a life preserver was your lying, cheating, good-for-nothing ex, nobody would blame you for taking it.

The first blitz against me was the AP hit job quickly following my senate testimony, but it certainly wouldn't be the last. Interestingly and with rare exception, the hit pieces on me, Paul, and the FLCCC weren't plastered across the front pages of major newspapers or featured on prime-time network news shows. Instead, they were largely placed in numerous smaller circulation, online magazines and news outlets. It must have been so stressful for the A-list editors and producers: "Man, I'd love to run these guys through the mud, but if we do, people might research them and discover they're legit! Let's let *Vice* cover it."

For the most part, the attackers left the FLCCC alone and focused on me or Paul personally. After all, it's harder—although not impossible—to take down an entire group of highly credentialed physicians and researchers than it is to lampoon a single individual. It only takes a whispered suggestion of impropriety, a twisted half-truth, or a fifteen-year-old accusation (not even conviction) of some minor offense to sow the seeds of doubt.

Fortunately, my colleagues and I have led perfect and exemplary lives absent of even a single error in judgment or behavior between us. (Whew!) Still, they targeted us personally and professionally, with headlines like this gem from Business Insider: "2 fringe doctors created the myth that ivermectin is a miracle cure for Covid-10 - whipping up false hope that could have deadly consequences."⁴ That one didn't just pummel me and Paul but also Paul's work on vitamin C. It was a true "two-fer" in terms of both man and medicine smearing.

The best line in that piece came just a few paragraphs in (all emphasis mine): "Together, Kory and Marik lead the Front Line Covid-19 Critical Care Alliance, a nonprofit organization founded by *fringe doctors* and former media pros that has led an increasingly *concerted campaign, hinged* on *twisted* science to *promote* ivermectin as a *cure-all* for Covid-19."

In case you didn't pick up on the subtly suggestive language there, *Business Insider* would very much like you to think that Paul and I are a couple of scheming, radical nutjobs.

"The Ivermectin Guys' Whole Thing Has Really Fallen Apart,"⁵ *Vice* declared. (Dammit, we never got the memo.) Never mind the fact that the headline sounds as if it were written by Bill or Ted of the *Excellent Adventure* series; in the first two sentences of the article, words used to describe us or our work include *vociferous, unproven, fringe, stood to gain money and attention, promoting that idea,* and *created a surprisingly durable little bubble.*

I'm not positive, but I think the "surprisingly durable" part may be a reference to Lucky Pierre.

ABC News ran a piece (later shared by *Good Morning America,* just in case they missed anyone) with this headline: "Ivermectin, condemned by experts as a Covid-19 treatment, continues to be accessible through telemedicine."⁶ In addition to recycling the same tired media narratives attacking the evidence base supporting ivermectin, they completely misrepresented my telehealth practice and made me out to be a money-grubbing physician preying on people for financial gain. So that was fun.

Scientific American took a tragi-comic approach with, "Fringe Doctors' Groups Promote Ivermectin for Covid Despite a Lack of Evidence."⁷ The subtitle: "The organizations touting unproved protocols for the antiparasitic drug may be harming vaccination efforts." Well, damn. I sure hope so.

The online magazine STAT, self-described as a "media company focused on finding and telling compelling stories about health, medicine, and scientific discovery," [note: they said *compelling,* not *true*] published this gem: "Encouraged by right-wing doctor groups, desperate patients turn to ivermectin for long Covid."⁸ Ridiculous political accusations aside, I suppose "right-wing" *is* more compelling than "fringe." (Apparently so did *HuffPost,* who took the heat off us doctors for four minutes with their expose on, "The Pharmacies Giving Ivermectin to People Bamboozled by Right Wing Misinformation."⁹ Bamboozled. Right-wing. Again.

If you wanted a one-stop-shop for trash talk on me, Paul, or the FLCCC, I'd send you to MedPage Today, the industry rag-mag that "normie" doctors love. Notorious for publishing Pharma PR pieces, they wrote an entire article based on the defamatory tweet by Kyle Sheldrick, one of the aforementioned ivermectin-fraud narrative promoters.

Probably my all-time favorite was their recently published hit piece[10] attacking the impending publication of the book you are reading. It unsurprisingly featured a quote from Dr. David Boulware, the assigned media attacker of HCQ and ivermectin. "Will serious doctors buy his book, no," Boulware insisted. "No one that's in the mainstream medical community will care what he has to say." Then they really pulled from the bank bench with a quote from Timothy Caulfield, LLM, a professor of health law and science policy at the University of Alberta in Edmonton with a history of attacking early Covid treatments.[11] "It is infuriating to see misinformation mongers being taken seriously," Caulfield huffed. "This kind of book can do great harm. It legitimizes bad science and erodes trust in scientific institutions." Actually, he kind of nailed it at the end as that is exactly what I am trying to do: *erode trust in captured scientific institutions.* He just left out the captured part.

If you were a Covid Dissident, a truth-teller, or a "right-wing fringe doc" with a big mouth and a public profile, you automatically had a target on your back. Just ask my colleagues and brothers in arms, Robert Malone, Peter McCullough, or Ryan Cole.

Based on the scope and scale of the attacks on Robert, combined with the (misguided) street cred of the publications running them, it was clear that he was the narrative's greatest threat. Adamantly and before most, Robert was warning the world about the dangers of the so-called vaccines. Robert is a physician, biochemist, and self-described "reformed academic" whose early work was instrumental in the development of the very mRNA technology he was criticizing as ineffective, toxic, and lethal. Robert was not a good look for Pharma.

So, they went big. *New York Times/Washington Post/Atlantic* kind of big. Even though it ran in an online nothing-blog called *For Better Science,* my favorite Malone headline of all time was this one: "How Dr. Robert Malone invented Antivaxxery." (I've yet to ask Bobby Kennedy his thoughts on that, but if I were him, I'd be pissed.)

I'm pleased to report that Robert is suing both the *New York Times* and the *Washington Post* for defamation for the ludicrous nonsense they made up about him in those articles. Living well may be the best revenge, but winning those two suits would rank pretty high on the list as well, I bet.

Peter McCullough has probably been trying to publicly disseminate Covid truths the longest. Peter is a cardiologist and internist (in addition to the former vice chief of internal medicine at Baylor University Medical Center and a professor at Texas A&M University) who came out swinging in early 2020, hitting hard at the insanity of not using hydroxychloroquine and other early therapeutics. He later championed ivermectin and appropriately called into question the safety and efficacy of the vaccines as soon as the data supporting them went sour. He was rewarded for his efforts with a near-daily deluge of credibility-crushing headlines such as, "Lawsuit: Doc Using Old Baylor Affiliation While Dishing COVID Vax Falsehoods,"[12] (good old MedPage Today).

And then there's Ryan Cole. Ryan is a dermatopathologist and the CEO and Medical Director of Cole Diagnostics, an independent medical lab. Ryan also made the mistake of throwing his hat in the ring in the Idaho governor's primary race, putting him even more sharply in the media crosshairs.

Ryan is one of the kindest, gentlest, and smartest people I have met on my Covid journey. I also frequently make jealous fun of him because he is a true Renaissance man. In addition to being a brilliant scientist, he is also an extremely talented woodworking artist with furniture pieces displayed in museums. Oh, he's also a skilled beekeeper, farmer, and livestock owner who makes and plays multiple instruments and is fluent in several languages. Obviously, I adore and respect the man . . . but the media certainly doesn't.

Ryan's been on the receiving end of the same media mudslinging campaign as the rest of us, with attacks including *false virus claims, rogue doctor, disciplinary sanctions,* and *treating patients 'beneath the standard of care.'* Again, attack the source and his message becomes moot.

Steve Christensen of Christensen Public Relations did an analysis of all the media mentions of me, Paul, the FLCCC, and ivermectin across TV, newspapers, online, and social media, and the results were fascinating. First, while there was an even split of positive and negative written articles about us, the positive reports were all on alternative or independent media and almost never in corporate controlled media. Second, he discovered an unprecedented discord between how ivermectin was discussed in print versus in social media. In the latter, the reports or mentions were overwhelmingly negative, lending credence to our belief that social media algorithms were designed to both censor and attack ivermectin.

I often wonder what would have happened if Elon Musk had bought Twitter early in the pandemic. I suspect things would have turned out quite differently for all of us.

Here's a quick recap of the career carnage caused by this bevy of blitzes:

Peter McCullough was fired from his position as vice chairman of internal medicine at Baylor University Medical Center and sued for violating their severance agreement. Baylor lost. Peter then took a job with a larger group practice. Two years later, they asked him to leave. Today he sees patients as an internist and cardiologist in private practice in Texas.[13]

Ryan Cole's private pathology practice, Cole Diagnostics, was dropped from two major insurance carriers, decimating his income, and forcing him to sell the practice. He estimates that this cost him future earnings over the next decade of approximately ten million dollars. Note he has six daughters, four of them currently enrolled in college. He also has complaints to his medical board for being a misinformationist.

The FLCCC's Umberto Meduri was forced to resign from his four-decade career working for the Veterans Administration, after he was threatened with the loss of his pension if he refused. His boss admitted that they "were under pressure from Washington, DC."

Paul Marik, as described earlier, underwent a brutal sham peer review process which resulted in the loss of clinical privileges at Sentara Norfolk General Hospital and the end of his clinical career. He no longer sees patients as he had a special medical license that was dependent on having hospital privileges.

Robert Malone and his wife, Jill Glasspool Malone, PhD, a scientist, physician, and vocal medical freedom advocate, pretty much lost everything to Covid. Most know Robert from his early career research in pioneering the biotechnology necessary to deliver mRNA into cells (which laid the foundation for the later development of mRNA vaccines), however his contributions to medicine, science, and the Dissident movement are truly innumerable. When he started to speak out publicly about the lack of vaccine safety and efficacy, Robert was repeatedly warned that his advocacy would harm the biotechnology consulting company that he and Jill had spent years building. The Malones refused to be silenced. One by one, their consulting and expert witness testimony contracts dried up. None of their countless contacts in industry or government would engage with them. At the time, Robert was a candidate for a highly salaried executive position with a company that was building a vaccine manufacturing plant in Egypt. That job offer disappeared. Their income has been obliterated.

Some of the most vicious attacks were launched at our FLCCC advisor from Brazil, Dr. Flavio Cadegiani, whom I consider the foremost Covid researcher and treatment expert in the world. To wit, he has published fifty-five papers, twenty-five of which have been peer reviewed and published with the rest still on preprint servers as of this writing. Further, he was the principal investigator of five double blind RCTS and fifteen observational studies. What is historically most important is his discovery that a class of drugs that suppresses the activity of testosterone led to better outcomes in hospitalized patients. His large RCT of the most powerful one, proxalutamide, was literally the highest quality rated RCT done in the pandemic. Hospitalized patients—even those being treated with ivermectin or nitazoxanide—were experiencing a 50 percent mortality rate. This was during the Gamma wave, the deadliest to date. With proxalutamide, that number plummeted to just 8 percent. Flavio cracked the code of the Gamma variant, saving countless lives.

The size, quality, and especially the findings of his study struck deep fear into the heart of Pharma.

To detail everything that happened to Flavio would require me to write the book *The War on Proxalutamide* (Jenna?), but the highlights are harrowing. His study papers were captured and held by multiple high-impact journals for well over a year before being rejected. He was eviscerated in the media and the *British Medical Journal* with false allegations of having committed "the worst ethical research violations in history." Alerted to possible threats against him, the Federal Police encouraged him to change his route to work and his clinic hours daily. He was threatened by two men in suits while sightseeing in New York City. Their words left no doubt that they had been spying on him; a listening device was later found in his home. A political party accused him of crimes against humanity before the international criminal court. Both his personal residence and his private clinic in Brazil were raided, during which documents and data associated with his scientific research were confiscated. He has been engaged in constant legal battles ever since, but as of the week this book is going to print, Flavio won the very last suit. An organization called Medicos Pela Vida ("doctors for life') wrote a glowing article[14] celebrating this fact with the headline, "Brazil's greatest living scientist is unanimously acquitted of all charges by medical councils." His private practice continues to thrive. Even after the hell he has endured, Flavio is one of the lucky ones.

In November 2021, I lost my third and final job in the pandemic. I was working as an independent contractor running the ICU in shifts at Aspirus Wausau Hospital Medical Center in central Wisconsin. I was initially hired by a small group of pulmonary and critical care doctors, but it quickly became clear that the chief medical officer and administration likely did not realize who I was when my paperwork made the HR rounds. Soon after being hired, I heard through the trusty hospital grapevine that the CMO was sending my colleagues a steady stream of bad press on me, trying to get rid of me.

At first, my new team defended me. "If Kory goes, we go," they told the CMO, and he backed down.

Then Biden's health-care worker vaccination mandate came down, and the CMO began harassing me about getting vaccinated. I kept deferring until the deadline loomed. When I announced that I had decided to get vaccinated and would be staying on (a lie to force their hand), the next morning I received a call from my ICU Director telling me that "he didn't need me anymore," mumbling something about how I had been reported for recommending against vaccination to a patient in the ER. It was pure BS and we both knew it.

"Sorry, Pierre," he told me. "This is a war, and unfortunately you're a casualty."

It was clear that I was now utterly unemployable by "the system." This event signaled the abrupt end to my ICU career.

I want to acknowledge that Paul and I are incredibly fortunate in that we are now employed by the FLCCC to continue our work and advocacy (wait until we address the War on IV vitamin C). But I have met countless doctors across the world whose livelihoods were destroyed without anything to fall back on. My colleague and early treatment advocate Dr. Andrea Stromezzi in Italy had his license suspended for twelve months, despite the fact not one of the 8,000 patients he treated had made a complaint. Andrea filed a suit with the Supreme Court which thankfully deferred his suspension. More recently, a pathologist colleague of his, Valerio Petterle, publicly revealed that he was seeing a large increase in vaccinated people "dying suddenly." For that, he was suspended from work for two months. Our dear colleague Jackie Stone in Zimbabwe, one of the earliest adopters of ivermectin in the world, literally saved thousands of lives. Her reward was several charges against her—one of which doesn't exist—by the Registrar of the Medical Council in Zimbabwe.[15] Possibly the most appalling punishment befell Swiss cardiologist Thomas Binder, whose social media posts criticizing the lack of science supporting Covid policies led to acquaintances reporting him because "they felt threatened." Police came to his home, and he was detained and forced to undergo psychiatric evaluation.

The censorship of me, the FLCCC, and our network of scientific and medical Dissidents trying to counter the narrative has cost millions of lives in the pandemic. History must document that. That is why so many of my colleagues are publishing books, in the hope of educating the masses.

Our path is not for the faint of heart—or their supporters. Months before this book's publication, a friend of mine in the UK posted a link to preorder it on Facebook. Her post was taken down within an hour and she was put in Facebook jail for thirty days.

In the 1930s, after the Nazi Party took control of the German government, Nazi students collected books they considered "un-German" and burned them in great public bonfires. Among the torched titles were the works of American author and political activist Helen Keller. In an impassioned letter published in the *New York Times* I once respected, Keller wrote, "History has taught you nothing if you think you can kill ideas. Tyrants have tried to do that often before, and the ideas have risen up in their might and destroyed them. You can burn my books and the books of the best minds in Europe, but the ideas in them have seeped through a million channels and will continue to quicken other minds."[16]

A radical, right-wing quack can dream.

CHAPTER THIRTY-FIVE

The Pharmacies and Hospitals Fall

In the wake of the global horse-dewormer propaganda campaign, hospitals started pulling ivermectin from their pharmacies. Health systems began harassing and threatening employees with loss of employment if they prescribed ivermectin. Pharmacies became even more brazen in their refusals to fill ivermectin prescriptions.

I'm running out of synonyms for the word *unprecedented*.

Physician friends were sharing memos they were receiving from their respective health systems. All of them read something like this:

"Due to the immediate news release from the AMA, APA, and ASHP, we are instructing all physicians to immediately stop prescribing ivermectin for Covid and Covid-related illness, as well as instructing pharmacists to stop dispensing. We must adhere to this mandate at once and cease all prescribing of ivermectin."

Hospital administrators, doctors, and pharmacists, "just following orders," stopped prescribing and distributing ivermectin. Online, an insider shared a festive holiday memo sent by his or her hospital's Chief Medical Officer, which read:

"Effective immediately, based on a vote by the PNT (Pharmacy and Therapeutics) Committee, Ivermectin *cannot* [emphasis theirs] be used for the treatment of COVID at [this hospital]. Currently, it is not on formulary and patients will *NOT* [emphasis theirs] be able to bring Ivermectin in as a "home medication." This decision aligns with evidence-based studies, the AMA, and the recommendations of our infectious disease specialists. *Our physicians and nurses in the COVID unit have had many patients request and, in some cases, demand Ivermectin* [emphasis mine]. We recognize this has posed many challenges and this official policy that [this hospital] does not use Ivermectin for the treatment of COVID will help to alleviate those challenges. Thank you and have a Happy Holiday."

Oh, there was a vote! At least they didn't just decide to withhold a lifesaving medication willy-nilly.

At the time the horse-dewormer campaign made its debut, I was successfully using ivermectin in high doses in all my ICU patients at the hospital in central Wisconsin where I worked. Within a few weeks, the Chief Medical Officer requested a meeting, where he informed me that our own P&T Committee would be deciding whether ivermectin should remain on the hospital formulary.

Ours was a smallish hospital, with around 200 beds including maybe thirty-two ICU beds. The head infectious disease doctor had initially been neutral on ivermectin but as we became closer as colleagues, I discovered he was open-minded, asked appropriate questions, and was receptive to receiving the data I had compiled. I began to share the mountains of evidence supporting ivermectin in Covid. He was shocked—and also became fully supportive of my use of ivermectin in the ICU, even going so far as to get the pharmacy to order more to support my clinical practice. It was incredible. He made it clear that his support was over the very vocal objections of the head infectious disease *pharmacist.* The latter, of course, was very much against the use of ivermectin in Covid, for the simple reason that his educational diet consisted solely of edicts pulled from captured high-impact medical journals and health agencies.

Fortunately, the head ID doctor outranked the pharmacist, so my patients' access to ivermectin was secure . . . until the ID doctor was suddenly reassigned to a different hospital in the system. Unfortunately, this happened just prior to the launch of the horse dewormer campaign.

Knowing that I was literally one of the world's experts on the clinical use of ivermectin in Covid, the CMO invited me to present to the P&T Committee in order to inform any decision on ivermectin that they would make. He told me that the infectious disease *pharmacist* felt it should be removed from the hospital formulary (of course he did), and that I could present my argument for why it should remain. Note that this is effectively what the P&T Committee in every hospital does: they decide on what therapies should be made available and for what conditions depending on their efficacy, risks, and especially, costs.

I immediately knew where this was heading, but if you know anything about me by now, it's that I don't go down without a fight.

The meeting was held on Zoom, and I pretended I was having a computer problem so that the pharmacist could speak while I supposedly worked out my technical issues. I wanted this shmuck to go first, to see what kind of butter knives he was bringing to this gun fight and so that I could counter optimally. It wasn't just a tactical play; I knew lives depended on this meeting.

The pharmacist began by presenting the first two of the Big Six fraudulent trials that had been published in high impact journals at the time, which I knew he was going to do. The rest of his slides consisted simply of the logo of an important health agency and their individual recommendation against ivermectin. WHO: against use outside of a clinical trial. Infectious Disease Society of America: against use outside of a clinical trial. European Medicines Agency, American Medical Association, Society of Emergency Medicine: against use outside of a clinical trial. On and on it went, slide after slide with the same recommendation. It wasn't a bad tactic actually as it really hit the point home.

Next, it was my turn. I began by detailing the many issues with those two trials (and they were glaring), and then reminded the group that the highest form of medical evidence is a meta-analysis. Then I presented the entire evidence base, including meta-analyses of the RCTs, the OCTs, and the health ministry data out of Mexico City, Peru, Brazil, India, and Argentina.

It was bulletproof if you ask me.

When I finished, I asked the committee if they had any questions. After a long silence—Del's NIH pause—a committee member asked, "Dr. Kory, why are your data and the recommendations of all the agencies so discordant?"

I also waited before replying.

"There is really only one answer to that, and that is that my group and I have no conflicts of interest around the use of ivermectin."

Mic drop.

They thanked me for my time, and I left the Zoom call so they could have their discussion and then hold their vote. Two days later, the CMO came to find me personally, sat down in my office, and informed me that the hospital had made their decision. They were removing ivermectin from the formulary and I would no longer be able to prescribe it to my patients.

As tragic and incomprehensible as that was, what was happening in the hospitals was nothing compared to what was happening "on the street." Retail pharmacists were increasingly refusing to fill valid prescriptions—something I had literally never witnessed in my entire medical career. Initially I fought back on the phone, citing my credentials and my research and explaining that I was one of the world's leading experts on the use of ivermectin in Covid. After a few early successes, it turned into a steady stream of losses. My despair was palpable and crushing.

It wasn't just happening in the US. A French colleague, Dr. Gerard Maudrux (the Pierre Kory of France), sent this disheartening note:

> A colleague was sentenced by the Council of the Order of Physicians to 18 months suspension for a prescription of ivermectin (I confirm, only one, in April 2020). The first sentence was 3 months, it was extended to 18 months on appeal. It was offered to him to declare that he had made a mistake, and that he would not do it again. As a result, his sentence was quashed. No, we are not among the Vietcongs [sic], in Burma, North Korea or China, but in France, the former country of human rights.

Early treatment doctors were forced to build lists of "safe haven" pharmacies where we knew we could easily get access to these medicines for our patients. Almost without exception, these pharmacies were independent compounding pharmacies. We were lucky here in the US because in many European countries, such a system of pharmacies does not exist. In Switzerland for instance, one physician told me the only way to get ivermectin was off the black market, where the price for a single 12 mg tablet was as much as 50 euro. This, to me, is the most convincing evidence of ivermectin's efficacy. Who would pay 50 euro for a tiny tablet that didn't work?

One night, I was inspired to make an attempt on a new, unknown pharmacy on behalf of a patient. I had just read a Substack by Steve Kirsch documenting another doctor's successful attempt at "swaying" a local pharmacy that had suddenly refused to fill ivermectin prescriptions. Steve's Substack included a letter[1] written on behalf of early Covid treatment pioneer Dr. Brian Tyson by his attorney. It was thorough, expertly argued, and served to inform such pharmacists that by refusing to fill valid prescriptions, they were: 1) violating patients' civil rights, 2) interfering with physicians' ability to practice medicine, and 3) exhibiting behavior that constitutes the unlicensed and negligent practice of medicine. Now, I had argued all these points before in previous "conflicts" with pharmacists, but never all at the same time, and rarely had I threatened legal action.

Duly and newly emboldened, I made the call. As soon as I said the word ivermectin, the pharmacist on the phone cut me off.

"I'm sorry but I cannot fill the ivermectin," he told me. "The pharmacy owner says there's no evidence it works in Covid, and we aren't to fill prescriptions for it."

I explained that he was the pharmacist on duty, and that I was a licensed medical professional calling in a prescription to him, not the pharmacy owner. He hemmed and hedged, and I cut him off. Holding Dr. Tyson's letter in my hands, I began spewing rapid fire arguments at him. I told him that my patient was a corporate executive and that his lawyer was prepared to send a letter of intent to sue if his prescription wasn't filled. I explained that by refusing to fill the prescription, he was violating my patient's civil rights, blocking my licensed ability to practice medicine, and clearly practicing medicine illegally and ignorantly.

"I am allowed to refuse," he told me.

"That may be what you think and what you have been told," I fired back. "But I can promise you that when you bring your arguments up in court, they will not hold up if any harm comes to my patient by your refusal. I'm sorry you're in the position you are in, but you have no rational or scientific evidence to support a refusal. If you want to go to court to find out, we can make that happen for you."

He continued to argue, and I continued to press. He told me he felt intimidated; I told him that was not my intention. I reiterated my arguments and asked for his name and license number. When he refused, I told him I was documenting the refusal. Finally, he relented.

"Tell me the rest of the prescriptions," he whispered.

I couldn't wait to call my patient and tell him that I'd won; I'd beaten the system. His meds would be ready for pickup within thirty minutes.

Only (and you probably saw this coming), they weren't. When his wife arrived at the pharmacy, she was told that her husband's ivermectin prescription would not be filled. Her husband threatened to sue the pharmacy as promised, but as far as I know they never went through with it.

Maybe pharmacists everywhere were simply too fearful to go against orders. Maybe most of them truly did believe the "ineffective horse dewormer" story and the Big Six trials they'd been fed. Maybe some of them rather enjoyed their newfound power over physicians they secretly resented or felt inferior to. Whatever the reason, the system had once again pitted an entire group of medical professionals against another, to the great detriment of sick and dying patients.

As my friend and fellow Covid expert Dr. Hector Carvallo long ago said, it was time for the lawyers.

CHAPTER THIRTY-SIX

A Legal Legend

The hospitals and pharmacies acted as if they were untouchable with their official, ironclad alphabet association orders in hand. What they weren't prepared for was a brush with Buffalo-born attorney Ralph Lorigo.

Ralph is staunch and savvy and, in my mind, one of the unsung heroes in the war on ivermectin. Starting in January of 2021, he began suing hospitals on behalf of the families of dying patients who were being refused ivermectin and other treatments in favor of the rigid protocol of an anemic dose of corticosteroids paired with the toxic, ineffective, and exorbitantly costly remdesivir.

Ralph's crusade began with an eighty-year-old woman named Judy Smentkiewicz, who'd been hospitalized on New Year's Day of 2021 and was in a coma. A family member saw my senate testimony and sent it to Judy's son Michael with a note: "Ivermectin! This is what Judy needs!" Judy's family pushed the hospital to give her ivermectin; the hospital pushed back. The family pushed harder; the hospital caved. Within twenty-four hours of a single 15 mg dose, Judy was off the ventilator.

But the hospital didn't continue the ivermectin and Judy soon began to regress. They moved her to a cardiac floor, where she was deteriorating quickly. The family demanded more ivermectin; this time, the hospital's *no* was unequivocal. Frustrated and furious, Michael retained Ralph Lorigo, who sued the hospital in an effort to get Judy the lifesaving drugs she needed.

The case was heard on an emergency basis as a literal "matter of life and death." The judge ruled in Judy's family's favor, and although at first the hospital refused to carry out the judge's order (can you *imagine?*), after a second hearing—while a patient was dying, mind you—they acquiesced.

Judy was put back on ivermectin and began improving almost immediately. Six days later, she was released from the hospital. Her story became the centerpiece of Michael Capuzzo's insanely compelling piece, "The Drug That Cracked Covid."[1]

Over the next year, Ralph took on over two-hundred cases in forty different states. In the first months of his legal ivermectin crusade, he was on a winning streak in court, getting judges to order the hospitals and doctors to prescribe ivermectin. In the vast majority of cases, the patients he fought for rapidly improved after being given ivermectin and were taken off ventilators and/or discharged. One man made a moving documentary of his plight to get his ventilated father ivermectin and posted it on Instagram.[2] I've seen it a dozen times and still can't watch it without tearing up.

Sadly, but not surprisingly, after a few successful months, the hospitals got wise. They began to fight back in court like they never had before, apparently terrified of allowing a precedent to be set which would force them to administer treatments demanded by patients or their family members that were contrary to the "expert advice" of their doctors. So, they began pulling every dastardly legal trick in the book, including appealing judgements while patients were dying on ventilators, then delaying or disobeying a judge's orders to give ivermectin.

In one case, Ralph requested that the judge find the hospital in contempt for refusing his orders, so the judge imposed a fine of $5,000 for every day the hospital withheld ivermectin. Want to know how long it took for the patient to be treated? A matter of hours.

Despite a string of court wins, Ralph found himself up against a new tactic. A hospital representative would claim in court that no physician employed by the hospital was comfortable writing the order for ivermectin. Ralph had to ask the judge to force the hospital to grant emergency hospital privileges for either a willing physician from the community or even the patient's primary care provider to write the order.

Of course, the hospitals delayed granting these privileges, and in some cases rejected the application of the physician as not meeting their privilege criteria.

It didn't stop there. In the cases where Ralph could find a doctor with privileges to write the order for ivermectin, the hospitals next told the judges that all of their nurses were refusing (or were told to refuse?) to administer ivermectin to the patient because they disagreed with the treatment. This forced the prescribing doctor to actually come into the hospital to administer it personally, often via nasogastric tubes if the patient was on a ventilator.

Enter Chicago physician and Covid superhero Dr. Alan Bain, a lovely and compassionate man who is expertly studied on ivermectin's efficacy. Despite Alan's already busy practice, he had become Ralph's go-to guy on the ground in Chicago, traveling miles a day between area hospitals, physically administering ivermectin to patients because the nurses were supposedly refusing.

How's this for a randomized controlled trial: Out of the 189 hospital ivermectin cases Ralph took on, eighty went to court. Of those, Ralph's team won forty and lost forty. Out of the forty cases they lost, thirty-nine patients died (97.5 percent). Out of the forty cases they won, only two died (5 percent). Even more specifically, Ralph went to court six times on behalf of patients hospitalized at Rochester Medical Center. In the three cases he won, all survived. In the three cases he lost, all three died. Of all the infuriating injustice I have witnessed in Covid, this one puts me over the edge.

Recently Ralph told me that, in his forty-eight years in practice, 2021 was the most satisfying in his career. As a long-time commercial litigator, he had never been able to directly save lives in court before; in that single year, he saved dozens, working tirelessly seven days a week. His last case was in January of 2022, after which Omicron took over and hospital cases plummeted. It was fortunate timing, as ivermectin was subsequently "proven" by the high-impact journal frauds to be ineffective.

Someone, somewhere, needs to make a movie about Ralph's life. His story may well be the *Schindler's List* of our lifetime.

CHAPTER THIRTY-SEVEN

Turning State's Evidence

In response to the insane pharmacy insurgency, a growing number of state attorneys general—including Nebraska, Louisiana, South Carolina and Oklahoma— took actions to protect physicians' ability to use off-label prescribing in the treatment of Covid. In an encouraging public statement, Oklahoma Attorney General John O'Connor said, "I stand behind doctors who believe it is in their patients' best interests to receive ivermectin and hydroxychloroquine."

Louisiana Attorney General Jeff Landry went after his state's Pharmacy Board for trying to block ivermectin access, writing in a letter circulated online, "A review of the US Food and Drug Administration's website found the agency acknowledges a physician may prescribe any FDA approved drug for off-label use for any number of reasons including that there may not be an approved drug to treat the disease or medical condition or that other treatments were not successful. Upon reviewing [the Louisiana Pharmacy Practice Act], I find nothing that would allow the Board to second guess the sound medical judgment of a physician when it comes to prescribing legal drugs to their patients, nor do I see anything that allows pharmacists generally to object to off-label use of FDA approved drugs."

It's shocking that it nearly took an act of congress to recognize this basic principle of medicine.

In an article in *Blaze News*, Landry didn't hold back. "I don't know where their conscience was when they were giving out opioids like M&Ms," he said. "Ivermectin is not even a scheduled drug. All of a sudden, they found a conscience?"

Boom.

In September 2021, the Texas Medical Board and State Board of Pharmacy issued a brief press release, just as the horse dewormer campaign was making the rounds. It stated, "The Texas Medical Board (TMB) and Texas State Board of Pharmacy (TSBP) do not endorse or prohibit any particular prescribed drugs or treatment for COVID-19 that meet the standard of care. Drugs are permitted to be prescribed off-label. It is the professional judgment of each physician to write their prescriptions while meeting all applicable federal and state statutes and rules. Similarly, each pharmacist must use their professional judgment in dispensing valid prescriptions while meeting all applicable federal and state statutes and rules."

It basically said, "Hey pharmacists. Let doctors be doctors. Unless you think you know better—in which case, be prepared to prove it."

In fighting back against federal regulatory capture, it is abundantly clear that the states must claw back their powers, which luckily already include the regulation of health care and medical practice. The problem is that few states asserted or exercised those protections in Covid, so it really depends on where you live. In my personal network of friends and colleagues who moved states during the last few years, this was the driving reason: to escape the places that were following the Fed's unjust, ignorant, and punitive regulatory totalitarianism. I know dozens who fled California not only for this reason but also because of the state's onerous and obsessive drive to force jabs on anything with a pulse, including all school-age children.

It does seem like a state rebellion is underway. On the Federation of State Medical Boards (FSMB) website, there is a color-coded legislative map of all the states with active bills regarding Covid legislation. In my cursory review, I'd say over 90 percent reflect legislation intended to preserve the autonomy and protect the licenses of physicians and pharmacists who use repurposed drugs.

Maybe medical integrity will turn out to be something red and blue can agree on after all.

CHAPTER THIRTY-EIGHT

The Miraculous Success of Uttar Pradesh

Author's note: Okay, so here's the deal. My brilliant, lovable, and expert cowriter Jenna McCarthy and I have been fighting about this chapter. A lot. It describes what I think is the most remarkable public health achievement in the history of mankind *and it deserves at least 100 pages to detail everything that happened so the world can learn a lesson, because these SOBs are going to keep springing lab concocted viral pandemics on us. But Jenna won't listen. She is insisting that I have to tell it in a few pages. So, I just want the world to know that this chapter was written under duress and in protest against her totalitarian ways. I hope you learn what you need to learn from this horrifically truncated description of what happened in Uttar Pradesh during Covid.*

Uttar Pradesh (UP) is a state in northern India that is roughly the same size as the UK and has a population of 231 million people. It's the home of the Taj Mahal and Hinduism's holiest city, Varanasi. If UP were a country, it would be the sixth most populous in the world.

What happened in Uttar Pradesh was the sole result of the integrity and actions of one man, Yogi Adityanath. Adityanath is a Hindu monk serving as the twenty-first and current Chief Minister of Uttar Pradesh who has a zero-tolerance policy against corruption. At twenty-six, he became the youngest member of Parliament in India's history, and in the first three and a half years after taking office, he took action against 775 corrupt officials in UP from the Indian Administrative Service and the Indian Police Service.[1]

His leadership in general—and in particular during COVID—should serve as a historically inspiring example to politicians around the world.

In March of 2020, Yogi Adityanath convened (and chaired throughout) a committee of eleven senior government officials tasked with managing different aspects of Covid. These included surveillance and contact tracing, testing and treatment, sanitization, containment, enforcement, doorstep delivery, issues of migrants, and communication strategy. The committee was widely known as "Team 11."

Team 11's Covid policies started out strong. Almost as soon as they convened, taking the lead from India's national protocol, UP quickly adopted hydroxychloroquine for use in prevention of Covid for all its health care workers as well as household contacts of all laboratory confirmed cases.[2]

Then, in August 2020, UP switched their protocol to ivermectin after an "experiment" in UP's Agra, a city of 1.6 million inhabitants. The head of the state's Rapid Response Team, Dr. Anshul Pareek, had decided to conduct a study of ivermectin as a preventive agent based on reports of its efficacy he had received.

Ten team members started with one pill every fifteen days. None got infected, even after coming into contact with many Covid positive patients.

Armed with this data, UP began administering ivermectin to close contacts of positive cases in the district to profoundly positive results. Based on these observations, the state health authorities gave the green light to use off-label ivermectin not only in prevention . . . *but in treatment.* They began to aggressively prophylax all close contacts of Covid patients and health-care workers while treating all patients with ivermectin. The *Indian Express* announced the switch from HCQ to ivermectin in an article published in early August 2020.[3] On August 28, 2020, the government of UP even tweeted out that the Department of Health would provide both HCQ and ivermectin.

It's worth noting that UP's government did what my colleagues and I had been imploring since the pandemic began: *They employed a risk/benefit decision-making analysis in an emergency.* Like you do in war. Even if the view was that the clinical trials evidence for HCQ or ivermectin was "insufficient," the evidence for harm was near nil, while the evidence for harm of widespread untreated Covid was obviously catastrophic.

As ivermectin was being distributed widely, "official" media mentions of its use became fewer and farther between. In December of 2020, *The Indian Express* ran a major article specifically highlighting the importance of ivermectin,[4] the last prior to the Delta wave in April 2021. So while Professor Paul Marik and the FLCCC are credited for the most public identification of ivermectin's effectiveness against SARS-CoV-2 by December of 2020, it should be remembered that the sixth most populous "country" in the world had already adopted its widespread use in prevention and treatment four months prior to my senate testimony.

Their protocol involved treating positive Covid cases with 12 mg of ivermectin for three days, and then reassessing based on response. It was real doctoring. They also used the drug in jails, where they reported that it cut the infection rate dramatically."[5]

Prior to the spring 2021 Delta wave, UP had only a tiny number of cases, despite massive testing (they ranked fifth in the world in this metric[6]) and despite being one of the poorest states in India.

In the first truly disturbing sign that global forces of censorship were being deployed, a month after the launch of UP's new ivermectin-based Covid program, the WHO posted a document called, "Learnings from Uttar Pradesh."[7] In it, the WHO glowingly detailed the comprehensiveness, sophistication, and resources invested by UP into their "test-track-isolate-treat," or TTT program. Neither the words *ivermectin* nor *treatment* are mentioned in the WHO document. Not even once.

By November, UP had the sixth lowest death rate in India (on the day of their program launch in August, they were tied for sixteenth). Two months later, deaths were virtually zero.

In March, the fierce new Delta variant was spreading in the adjacent Punjab region of India; next Delhi was hammered, experiencing 50 percent higher rates of death than its nearest neighboring city. Both Delhi and Mumbai implemented harsh lockdowns, causing massive numbers of migrant workers to flee to their hometowns; for many, that was Uttar Pradesh.

India was a mess, particularly the big cities like Delhi and Mumbai. Hospitals were overrun and running out of supplies, sparking reports of the collapse of city and regional health systems.

Uttar Pradesh was not spared. They went from 300 cases a day on March 19th (*out of 231 million people*) to 2,589 just two weeks later, and then to almost 40,000 by late April. Note that the United States of Pharma would have loved to see "just" 40,000 cases a day.

The massive surge was met with an aggressive response by UP's Team 11. Keep in mind, they had already been distributing ivermectin doses bi-weekly to over one million health care workers since August 2020.[8] They then deployed 400,000 health care workers to more aggressively perform testing throughout the state and began screening all incoming migrants at bus stations, airports, and train stations.[9] By May 15th, they had conducted 43 million tests.[10] Teams proactively visited homes, testing Covid-symptomatic individuals using Rapid Antigen Test (RAT) kits; those who tested positive were isolated and given a medicine kit containing ivermectin with clear-cut treatment instructions.

With millions of migrant workers bringing Delta into UP, the result was one of the highest infection rates in India. Notably, despite this, their death rate was literally one of the lowest.

How did they achieve the latter? On April 17th, UP's government released a list of seven medicines, published in a major newspaper, giving clear instructions on how to treat patients with Covid.[11] In particular, they advised giving ivermectin after food, which we now know leads to much higher efficacy. They also included guidance about drinking enough water and getting enough sleep. Imagine that.

I received an email from a surgeon and the owner of a hospital in India who asked to remain anonymous. These are his words, which I am transcribing directly:

> I am treating [a] lot of people through phone consultations. My policy is "Pill First and Test Next." We are getting excellent results with ivermec-tin (started early), BIG Doctors who constantly appears on TV don't want to give credit to this cheap drug. None of my hospital staff got positive till now. Nothing exaggeration. We still believe that IVERMECTIN is SAVING INDIA. The most powerful weapon of India (IVERMECTIN) has been getting BAD propaganda by Big pharma & Big scientists since Jan 2021 (Coincidentally the month when global political power change). Many Indian doctors under the influence of their NRI doctors have stopped using ivermectin. This lead to neglected early home-care and people presented to hospitals in late state.

His email went on to explain that the use of high flow nasal cannulas was causing an unexpected oxygen shortage, and due to the scarcity of both oxygen and hospital beds, the Indian government was finally allowing iver-mectin for early home-care treatment.

"This is a GAME CHANGER decision for INDIA and even for the entire WORLD," he wrote. "Ivermectin saved India in 2020 after it got official permission in Uttar Pradesh. This led to control of Covid by October 020. Now with the official acceptance of Ivermectin for Covid-19 . . . we are sure that the steep spike will have an ABRUPT FALL in 2 weeks. It's not a prediction, but a reality. Trust Ivermectin. It works on ALL VARIANTS of Coronavirus. We BEG health agencies & mainstream media in other countries, NOT to give BAD PROPAGANDA to Ivermectin (a SAFE HUMAN drug since 33 years). Ivermectin is saving India."

When I replied with an invitation to join our FLCCC webinar to discuss the situation in India, he replied:

I have discussed about your webinar with my government doctors. They said not to participate in webinars about drugs not endorsed by WHO. But keep updating you through FB & WhatsApp. Australian MP Craig Kelly's FB was also deleted permanently 3 days back, because of IVERMECTIN posts.

I have had the pleasure of meeting and working with former Australian Member of Parliament Craig Kelly. He was an aware and vocal proponent of successful early treatments like ivermectin and HCQ and like Senator Ron Johnson, led a very public and much-vilified attempt to challenge Australia's criminal denial of early treatment. In a tweet, he humorously asked if Australia could borrow UP's Chief Minister Yogi Adityanath to replace their "hopelessly incompetent State Premiers." The Chief Minister's office replied, "We would be happy to host you & share best practices which helped #UttarPradesh fight the pandemic under the guidance of Hon. PM Shiri @narendramodi Ji & leadership of #UPCM Shiri @myogiadityanath Ji. Let us collaborate & co-operate in this global fight against #COVID19."

In another shining example of pandemic leadership, Adityanath announced that the UP government would bear the cost of Covid treatments, even in private hospitals.[12] The Rapid Response Teams spread out across the state, visiting 97,000 villages, testing widely, prophylaxing close contacts, and *treating the Covid-positive with ivermectin.*

Cases started to drop precipitously. Based on data from Johns Hopkins University, on April 26, UP had 33,531 cases, which dropped to 18,023 by May 12. By May 30, cases had dropped to under 600 a day. Again, a state with 231 million people that was performing massive testing saw fewer than 600 cases a day. While the rest of India was still raging with Delta.

On May 12, one of the last headlines celebrating ivermectin use in UP was featured in a major newspaper, *The Indian Express*.[13] Nine days later, the *Hindustan Times* reported on the amazing turnaround in UP, detailing the steep drop in cases, the high recovery rate of cases, and one of the lowest positivity rates in the country.[14] *They did not mention ivermectin once.* Instead, they reported on the number of vaccines administered in the state. It was absolute nonsense, and the first hint of what would become a herculean attempt to credit UP's turnaround to vaccination. (For reference, the *Times of India* reported in August of that year, more than two months later, that just 5.8 percent of the UP population was fully vaccinated.)[15]

In the meantime, major US media was focused solely on UP's massive infection rates and widespread ivermectin use to support their "ivermectin doesn't work" narrative. Because UP had so rapidly and successfully extinguished cases and deaths, the media avoided mentioning that UP was using ivermectin systematically in prevention and treatment.

Once again, the WHO posted a fawning review of UP's Covid response; once again, ivermectin did not appear anywhere in the report.[16] Instead the article exclusively credited UP's testing and contact surveillance methods for its success. The piece included this hilarious throwaway line, "*Those who test positive are quickly isolated and given a medicine kit with advice on disease management.*" Similarly, India.com announced that Yogi Adityanath's government was preparing five million "special medicine kits" for children with chewable tablets and/or syrup.[17] *Nowhere is it mentioned what the "special" medicine is.*

It's probably not important what was in those kits. (Eye roll.)

Case counts dropped precipitously. Team-11 and its massive workforce continued testing, tracking, and treating. By the end of the summer, *Covid was effectively eradicated in Uttar Pradesh, a feat I consider to be one of the greatest public health achievements in history.* One of history's most highly contagious, aerosol-transmitted viruses had essentially disappeared from within the borders of a massive state of 231 million people.

On September 10, the *Hindustan Times* reported jaw-dropping findings that should have made global headlines.[18] First, sixty-seven of UP's seventy-five districts had not reported a new case in the previous twenty-four hours. (That would be like forty-four states in the US not reporting a new case at the same time. Think about that for a second.) In thirty-three districts, there was not a single "active case." In fact, in the entire state of 231 million people, there were only 199 active cases. (The USOP has more monkeypox now than Uttar Pradesh has had COVID since, for pathetic reference.) In the previous twenty-four-hour period, 226,000 tests had been conducted, resulting in eleven new Covid-positive cases. That's a 0.004 percent positivity rate, incidentally. Effectively zero. In contrast, Kerala, one of the two states in India that rejected the use of ivermectin, was reporting a 19.7 percent positivity rate at the time.[19]

This astonishing news should have been on the cover of every major newspaper across the world. UP's TTT program had ended the pandemic within the state. It was akin to discovering another penicillin, and yet the media blackout was global. Once again, any mention of UP's success was roundly and ludicrously attributed to the vaccines, even though by September 21, still only 10 percent of the country was fully vaccinated.

In one final blow to ivermectin's credibility and likely future, a wickedly detailed, comprehensive 132-page report on UP's TTT program was published in October of 2021, a month after they had eradicated Covid. Written by a professor at one of India's top universities, the report mentions ivermectin exactly once—as a "protocol medication they monitor the supply of."[20]

The United States of Pharma had won. Ivermectin's efficacy was buried by mass censorship. I wouldn't be surprised if some veterinarians even stopped using it to treat horses with worms.

CHAPTER THIRTY-NINE

The WHO Goes After India

In the rest of the country outside of Uttar Pradesh, India's federal government started off on the right foot. In March 2020, they issued a recommendation for the use of HCQ as a preventative for health care workers across the country.[1] The Indian Council of Medical Research (ICMR) quickly conducted a trial and found that HCQ reduced infection rates by up to 80 percent in these workers.[2] The prestigious All India Institute of Medical Science (AIIMS) even suggested its use in treatment while also indicating that ivermectin could be used as an alternative.

As a nation, India ramped up their production of HCQ and, by February of 2021, just before Delta hit, they had distributed over 100 million tablets across the country.[3] They also exported HCQ to 97 countries around the world.[4]

As Delta began wreaking havoc in India, their federal agencies again set a global example. On April 22, 2021, the AIIMS issued new guidelines which now included ivermectin in their "may do" category. Less than a week later, the Ministry of Health, the AIIMS, and the ICMR updated their national COVID-19 treatment protocol. The new protocol *recommended ivermectin and budesonide for all patients with a mild case of COVID*. It wasn't exactly what I would have done—why limit it to mild cases?—but it was a start, and they even recommended it for three to five days.[5]

For the enemies of ivermectin (of which there are many), this was not good news. The world's second most populous country was recommending ivermectin to over a billion people. Clearly, something would need to be done to stop this.

Indian media didn't touch the story, instead continuing to promote remdesivir as an effective treatment. It was as if there were two different realities—in the local health systems, millions of patients were receiving ivermectin, yet the general public knew nothing about it due to a near-complete media blacksout. The few outlets that even bothered to mention ivermectin only did so in the context of "an unproven or outdated treatment."

Complete control of national media wasn't enough. A random branch of India's federal health system, the Directorate General of Health Services (DGHS), issued a bulletin "not recommending ivermectin." One problem: the DGHS isn't a governing body, but a "repository of technical knowledge." They had no authority to impact national guidelines.

Nevertheless, the DGHS bulletin was picked up by the media (of course) and caused widespread confusion. They—presumably the WHO and BMGF—were working overtime to create a narrative that India was reversing its stance on ivermectin.

Here's where things took a turn worthy of a Tarantino film: Two weeks after the Indian feds included ivermectin in their national treatment guideline, the WHO made a move. On May 8, the WHO's chief scientist Soumya Swaminathan (an Indian, for crying out loud) posted the organization's laughable "Mild Covid-19 home care bundle" to Twitter.[6] This cutting-edge protocol included *isolation, hydration, oxygen monitoring, and Tylenol*. It contained no actual treatment, just like the NIH's guidance issued in the US.

Three days later, Swaminathan attacked ivermectin directly in another tweet contradicting India's federal health policy, writing, "Safety and efficacy are important when using any drug for a new indication. @WHO recommends against the use of ivermectin for #COVID19 except within clinical trials."[7] She foolishly included a link to Merck's statement in her tweet, literally citing a pharmaceutical company's public relations campaign against a competing drug. You cannot make this stuff up.

Swaminathan soon found herself in deep doo-doo. The Indian Bar Association swiftly filed criminal charges against her for this tweet, accusing her of misconduct by misguiding the people of India to further the agenda of special interests and to maintain an EUA for the lucrative vaccine industry—a crime which included the possibility of a death penalty.[8] She quickly deleted her tweet.

An Indian physician and follower of the FLCCC emailed me during the Delta wave. "Unfortunately, the WHO Recommendation and the *JAMA* study have done more harm than good," he wrote. "We doctors who believe in ivermectin have been using it even today but the ones who were a little skeptical, might have stopped thanks to the misleading information lately. There are absolutely no public information initiatives on ivermectin. I have a voice but not loud enough to reach everyone. The kind of popularity remdesivir has got [sic] I wish ivermectin did and we would have been at a different place altogether."

You can say that again.

By the end of September 2021, most of India was out of the Covid woods due to abundant natural immunity and widespread use of ivermectin. Naturally, this is when our buddy "Gilly Bates" (what I like to call him) paid a little visit to India to meet with Prime Minister Narendra Modi. Reports of the visit featured a giddy Gates praising the launch of the government's creation of a digital health infrastructure (ahem) that would "help ensure equitable, accessible health-care delivery and accelerate progress on India's health goals."[9] Or something like that.

Curiously (coincidentally?), ivermectin and HCQ were both dropped from India's national guidelines two days after this visit.[10]

To any naysayers and propagandists who are skeptical of the public health data from Uttar Pradesh in particular or India in general, similar programs were launched in Paraguay,[11] Argentina,[12,13] Brazil,[14] Mexico,[15] the Philippines, and Peru.[16] Although none came close to the scope and scale of UP's program, all saw astonishing success.

What happened in India was tragic on every conceivable level. It wasn't merely a matter of "not knowing how to treat a novel virus." They had essentially figured it out, and seeing that they had, the wicked powers-that-be staged a deadly, coordinated coup. I pray daily that I live to see justice served.

CHAPTER FORTY

The Vaccine Disinformation Campaign

No book detailing the widespread fraud, corruption, and propaganda that marked this three-year period in history would be complete without at least touching on the vaccine madness. While clearly this topic could fill several books of its own, after detailing the highlight reel of the disinformation campaign against ivermectin within the medical community, I want to address the similar campaign around the vaccines that was occurring at the same time, and which continues to this day.

I had been tipped off to alarming toxicity signals associated with the Covid mRNA vaccines through the VAERS database within weeks of the rollout, but I began scrutinizing the data (not) supporting the safety or efficacy of the mRNA vaccines as well as Novavax a few months later. I did not yet know a vaccine injured person, as most around me seemed to be getting vaccinated without issue. I found out later this was not true in many cases.

As my developing research into the Disinformation tactics against ivermectin science was growing, I saw *the exact same tactics being deployed around the vaccine campaign*. It was the ivermectin nightmare on a nonstop loop.

From the initial trials, deceitful actions such as unblinding trial subjects, burying evidence of infections occurring by miscategorizing them, not following or appropriately documenting subjects with adverse events, manipulating patient files, removing subjects in order to hide serious adverse events, and publishing trials overstating vaccine efficacy were occurring with startling frequency.

Regarding that last part—overstating vaccine efficacy—I'm compelled to share a quick lesson in the ambiguity known as "efficacy." With the majority of the Covid vaccines, you probably heard (over and over ad nauseam) that they were 95 percent effective. Sounds impressive, right? *These vaccines prevent infection in 95 percent of recipients!* The problem is that isn't at all what that means—and nobody bothered to explain that to the world. The 95 percent referred to what's known as *relative risk reduction* (RRR), when in fact what matters is *absolute risk reduction (ARR),* which turned out to be . . . wait for it . . . *less than one percent.*

According to a report in the *New England Journal of Medicine,*[1] in an early Pfizer vaccine trial of 37,000 subjects, 170 developed Covid. Eight of them were in the vaccine group and 162 were in the placebo group. Infection rates were therefore 0.04 percent vs. 0.88 percent, a *relative* efficacy of over 95 percent. The *absolute* difference between groups was 0.84 percent, meaning that in this trial, the vaccine prevented one Covid infection for every 119 people vaccinated—or less than one percent. The Moderna, AstraZeneca, and Johnson & Johnson "vaccines" have shown similar results.

I ask you: How many people would have lined up for an experimental jab (side effects notwithstanding) if they knew that it reduced their odds of catching the dreaded virus by a lousy one percent?

Beyond the original vaccine trials, I began to observe and document the retracting of papers detailing adverse events or inefficacy. Just this week (spring 2023), I learned the journal *BMC Infectious Diseases* will be retracting a paper by Mark Skidmore, a highly respected economics professor from Michigan State University. In that paper, Mark's analysis found that up to 280,000 US deaths were related to the vaccines.[2]

Just as with ivermectin science, with extraordinarily rare and only recent exceptions, journal editors have systematically rejected any study which accurately documents the inefficacy or even negative efficacy of the vaccines. And side effects? What side effects? "Safe and effective," folks! Of course, with the exception of those *very, very rare* and *very, very mild* anomalies.

I saw severely flawed (fraudulent) mathematical analyses which concluded that millions of lives were saved by these so-called vaccines, papers which were trumpeted across corporate-controlled and social media. I witnessed firsthand how vaccinated patients were routinely not documented as vaccinated when being newly admitted to hospitals for Covid in the US. I quickly discovered that the proportion of hospitalized patients the CDC was showing as being vaccinated was extremely and falsely low, and severely discordant from numerous foreign national health agencies which were not only more transparent, but far more accurate in the documentation of vaccination status of hospitalized patients. Within six months of the rollout, it was clear to me that the rates of hospitalization of the vaccinated exceeded those of the unvaccinated, yet the boosting and cheering and coercing and maligning continued apace, using numerous false narratives.

I have watched major media, again with scant exceptions, consistently fail to cover the highly disturbing explosion in life insurance claims and excess mortality in young people between the ages of eighteen and sixty-four, all beginning after the vaccine rollout. I've read hundreds of newspaper reports of young, otherwise healthy people doing activities they love dying suddenly and mysteriously all over the world. I've seen celebrities, athletes, musicians, reporters, TV broadcasters, and comedians keel over (sometimes fatally) *right on camera*. Never is the vaccine mentioned as a possible cause.

I have seen baseless claims that the vaccines are indisputably safe in pregnancy, based on numerous peer-reviewed and published papers using willfully manipulated analyses purportedly showing no risk of harm. I have witnessed the systematic ignoring of toxicity signals in pregnancy screaming from VAERS and all the other foreign national vaccine adverse event databases, not to mention massive social media groups of women being shut down for discussing the alarming rates of miscarriages and menstrual irregularities. I have heard nothing but deafening media silence on the unprecedented month to month drops in birth rates[3] from countries across the world, beginning almost exactly nine months from the ramp up of their respective vaccine rollouts.

There was also a new dynamic within the case reports of severe adverse events. In any report detailing some horrific outcome or adverse event, sentences were inserted which claimed that these events should be considered "very" (not a scientifically defined term) rare, and then reminding the reader that the vaccines are safe and effective. Aaron Hertz is a friend and brilliant rabbinical scholar and the author of the Substack, "Resisting the Intellectual Illiterati." He has taken it upon himself to compile every published case report of an adverse event associated with these vaccines. As of today, he has a collection of over 3,000 reports of dreadful adverse events associated with the vaccines. These aren't VAERS entries; they are detailed analyses published in peer-reviewed journals. It's an astronomical number.

Another friend and colleague, Ed Dowd, a talented and tenacious Wall Street analyst and the author of *Cause Unknown: The Epidemic of Sudden Deaths in 2021 and 2022* (Skyhorse Publishing, 2022), assembled a team that analyzed life insurance, epidemiologic, and disability data collected during this time period. The resulting conclusion is unequivocal: we are experiencing an unprecedented humanitarian crisis—one the media is categorically ignoring. Young people are dying or becoming disabled across the world at rates that have never been seen. Excess mortality continues to rise, despite the current mildness of Omicron and its subvariants. Meanwhile, you can't hear yourself think over the media's, journals', and our health agencies' deafening chorus of "safe and effective" with regard to the vaccine . . . not to mention somehow necessary for everyone, even toddlers, children, young adults, pregnant women, older people, and inexplicably those with natural immunity, an absurdity never before even considered. It's a relentless and unwavering narrative identical to that of ivermectin being a worthless horse dewormer.

I am now estranged from not only those who practice medicine inside that system, but from "science" in general, at least as it has come to be known. I no longer know who and what to trust within that system, and have for now, chosen to believe nothing that cannot be confirmed by numerous objective data sources using an assessment of "the totality of the evidence" and not the curated, predetermined conclusions found in the high-impact medical journal studies. To say it is a sad state of affairs is the understatement of my life; to realize that this state of medical science has existed for decades is both humbling and terrifying. How many people have I hurt using medicines built on lies in my career?

My consolation is that oftentimes, it takes great destruction to realize where weakness lies. After a natural disaster leaves a community in ruins, you can bet the rebuilt structures will be engineered to withstand the next one. At least I know what I'm dealing with now, because only good things can come from that knowledge. In one film depiction of Pearl Harbor, the attack's planner, Japanese Admiral Isoroku Yamamoto, declares, "I fear all we have done is to awaken a sleeping giant and fill him with a terrible resolve."

Yes, I am awake . . . and I am filled with a terrible, immense, galvanizing resolve.

CHAPTER FORTY-ONE

Rallying the Troops

By late 2021, the Dissidents—and scores of Americans—had had enough with the masks, mandates, lockdowns, and illegal governmental overreach. People were protesting around the world, and it was about time the red, white, and blue got in the game.

The FLCCC was honored to help plan and fund the two national Defeat the Mandates Rallies in LA and DC. These massive events were organized in just three months by the incredibly committed and indefatigable Matt Tune and Louisa Clary, with support and direction from Steve Kirsch and his VSRF. Matt is a military veteran and NBC executive whose job was being threatened by vaccine mandates; Louisa is a former real estate investor, an early and vocal ivermectin advocate, and a prolific Covid social media poster I began following early in the pandemic who later became the Executive Director of the VSRF. Together, Matt and Louisa made a formidable team.

With the first Defeat the Mandates rally, the biggest challenge was getting the word out in such a short period of time. How would thousands even learn about it, much less attend, when social media was censoring anyone and anything anti-narrative, and mainstream media was demonizing anyone who opposed our oppressors?

Once again, the answer was Joe Rogan.

Louisa and Matt knew that Dr. Peter McCullough was going to be a guest on *The Joe Rogan Experience* in mid-December 2021. Peter agreed to plug the DC rally, which was planned for the following month. Peter's interview went viral, which would have been amazing had it not been promptly taken down by YouTube.

Desperate to continue the momentum, Matt and Louisa wrote to Rogan, requesting he interview Dr. Robert Malone. While it typically takes weeks or months to get on his calendar, Rogan and his team understood the urgency of promoting the DC rally. Joe came in on his day off to record Robert just six days later, on December 30th. Rogan's team edited and released the podcast the very next day, on New Year's Eve. Robert's endorsement of the Defeat the Mandates rally was a game-changer. The day before recording that podcast, Robert had been deplatformed from Twitter—and people were pissed. A firestorm was ignited, and views and downloads of the Rogan podcast skyrocketed.

Epic sidenote: In a fortuitous stroke of luck, Robert happens to bear an uncanny resemblance to actor Jonathan Goldsmith, Dos Equis beer's "Most Interesting Man in the World." A meme of with Robert's face and the quip, "I don't always lose a half a million followers on Twitter, but when I do, I gain 50 million views on Rogan" went viral.

By January 1, YouTube had already taken Robert's episode down.[1] In the interview, he used the term "mass formation psychosis" to explain the seemingly inexplicable loyalty of a wide swath of the global population to absolutely asinine beliefs around Covid. If you searched for the phrase at the time, you'd get Google's new Covid-era auto-reply, "the results are changing quickly."[2,3] It was the internet equivalent of, "I'm sorry, Simon isn't in right now, may I take a message?" *when Simon was sitting right there.*

Millions learned about the DC rally from Peter's and Robert's appearances on Joe Rogan. Video clips of their interviews spread like herpes across social media. Despite being heavily patrolled by the Facebook censorship police, a Defeat the Mandates Facebook group sprang up and quickly gained thousands of followers. Thanks to this exposure, thousands of people showed up at the Lincoln Memorial on a cold day in January—they were vaccinated and unvaccinated; Republicans, Democrats, and independents—to stand up for freedom.

And to think, it nearly didn't happen. Between massive operational costs, an impossible timeline, and the city's barrage of Covid restrictions, the organizers faced a nonstop parade of planning challenges. In late December, DC Mayor Muriel Bowser coincidentally re-instated the mask mandate and put other restrictive, city-wide directives in place effective January 15—one week before the rally. Bowser's lovely new decree required anyone age twelve and older to have a Covid vaccine in order to enter indoor venues including restaurants, bars, and gyms, or even to stay in a hotel.[4]

How would unvaccinated people attending the rally buy food? Would they be allowed to use public restrooms? Where would they stay? Would they be arrested for not wearing masks? Two weeks before the event, a late-night meeting was called among the leaders of the collaborating organizations to discuss these challenges. All of the unknowns left many on the planning committee uneasy. Louisa and Matt were criticized as not being cautious and calculated enough with their planning; some were predicting failure. It was a critical moment where several funding the effort threatened to back out. I wasn't about to let the whole thing go up in flames. Not after how hard we had all worked—especially Louisa and Matt.

"We are at war," I said. "And you go to war with the army you have, not the army you want [I was referring to resources, not Matt and Louisa]. We have to take risks. The time is now. I'm in." With compelling vocal support by Michael Kane, a community organizer extraordinaire who works for Children's Health Defense, the group acquiesced.

Under Louisa and Matt's organizational leadership and with the help of Amanda Damian, a logistics superstar with an unmatched dedication to the movement due to her own Covid vaccine injury, a group of some of the best people and organizations in the medical freedom movement successfully pulled off the DC and LA rallies. Notable rockstars included Del Bigtree and his team at The HighWire and ICAN; Bobby Kennedy, Laura Bono, Michael Kane, Stephanie Locricchio, and Aimee Villella from Children's Health Defense (CHD); Richard Urso and Ryan Cole from the Global Covid Summit; Jeff Hanson and Laura Sextro of the Unity Project; FreedomMed; my incredible team at the FLCCC and many others.

"It's no business of the federal government or agencies to tell doctors how to practice medicine," Paul declared in his speech at the LA rally.

I added my own sentiments. "The world has gone mad," I insisted, "and it is not their fault. It is from unrelenting propaganda and censorship. We live, unfortunately, in the United States of Pharma. They do not care. They put profits ahead of people's lives. They do this rapaciously, relentlessly, and they are doing it today. And nobody is aware."

The cost of the rallies was almost as insane as the speed of planning. Nearly a million dollars was spent on each event, funded not only by our respective organizations but tons of small donors as well. After each rally, we were left with sizable money shortfalls. Louisa and Matt, after having worked around the clock for months, had to double down and scramble to raise the necessary funds *after* the events. But they got it done, and the impacts were profound.

I will never forget standing on the steps of the Lincoln Memorial with Paul, looking out over a massive sea of people screaming their support, many holding FLCCC signs and wearing FLCCC gear. Talk about validation. It was a powerful and much-needed reminder of what we were fighting so hard for.

The People's Convoy of truckers was another unforgettable experience. In February of 2022, more than 1,000 semi-trucks, RVs and cars drove from Barstow, California to Washington, DC to protest the sweeping Covid restrictions being imposed across the country. Everyone involved did so at immense sacrifice, pausing their ability to earn money to support their families and sleeping in their cars and trucks along the way, many with their families in tow. *In February*. Millions of people came out in support along the highways and overpasses to cheer the Convoy on with signs and American flags. If you never heard about it, I'm not surprised. It was largely ignored by mainstream media, who were very busy trying to drum up sympathy and support for the war in Ukraine. (One notable exception was an article in *Forbes* shaming the People's Convoy organizers for—and you cannot make this up—bemoaning Americans' loss of freedoms, which *were nothing compared to what Ukrainians were experiencing*.)[5] When the convoy got to DC, Senators Ron Johnson and Ted Cruz invited the convoy organizers and some participants to a press conference to explain what the Convoy was about and what we were trying to accomplish. You can watch it on Rumble[6] and if you get a chance, I highly recommend it. There are many powerful and moving moments (as well as more than a few heated replies—mine included—to rude, insulting questions posed by the smug, so-called press).

The energy and excitement and community and connection of those events were unparalleled experiences for our group of Covid activists. While it's hard to measure the direct impact that the rallies or the convoy had on Covid mandates, the fact that the mainstream media couldn't entirely ignore the thousands of passionate supporters that these events attracted was a win in and of itself. The *New York Times* even put the DC rally on their front page. These events opened up a dialogue among Americans in all walks of life, creating incomparable support for health freedom organizations and standing as pivotal moments in defining a movement across the US.

CHAPTER FORTY-TWO

Covid Highs and Lows

When I look back over my calendar entries from the past eighteen months, the number of lectures, events, and media engagements I have participated in is truly staggering. I was willingly running myself ragged for one reason: if you recall, since the beginning of my career, I have found it hard to say no to any professional request, whether it's an invitation to lecture or collaborate, or if a colleague simply wanted me to help out with an initiative or research. I can recall conversations with my wife fifteen years ago where she would counsel me on how to say no when she could see I was burned out. These overly busy episodes would come in waves throughout my career, but none can even begin to compare to the past three years. My hope is that finishing this book will leave a hole in my schedule that I will not immediately fill with a string of work tasks but instead can give to my family and my health.

There have been many high points in my Covid journey—and also more than a few lows, including losing three jobs and having to say goodbye to my ICU career. I'm sure I've forgotten several of both, but here are some of the best and worst moments that stick out.

At the height of the frenzy of activity, beyond the repeated media hit jobs and the PR crises every time one of the Big Six was published, the attacks started to get personal. By the end of 2022, I had received eight separate complaints to the Wisconsin Medical Board, either accusing me of misinformation or requesting that my license to practice medicine be revoked. One complainant wrote that I "bring disrepute to the physicians of Wisconsin."

Ouch.

I have answered each of the accusations with mountains of data supporting my position but have yet to hear back from the board. Luckily, they seem to be extremely busy or just poorly resourced. I know many physicians from other states who have not fared so well, with either immediate suspensions or revocations of their licenses.

May of 2022 saw a record low when Paul Marik, Peter McCullough, and I each received letters from the American Board of Internal Medicine (ABIM) informing us that we had run afoul of their shiny new "misinformation policy," and threatening to decertify us. Traditionally, a board certification is not a license to practice medicine; it is (or was intended to be) a badge of distinction for passing a more rigorous certification exam. The problem is, over the past decade, many insurance companies require board certification to be on their panels, and many hospitals require it to join their staff. So, losing this status can have a real impact on a physician's livelihood. Or it could have if I planned to return to that system—which I do not. My current practice is fee based and private.

Still, we decided to not go down without a fight. We answered their complaints with a massive rebuttal letter loaded with references. It's been months now and Peter is the only one of us who has gotten a reply. And do you want to know what their response to his submission of all the data showing the toxicity and lethality of the vaccines was? They informed him that he was wrong, based on the fact that the CDC had determined the vaccines were "safe and effective." I am not making this up. That was literally their entire argument.

Hovering in the same vicinity as the ABIM heartache on my personal Covid lowlight list was when California passed Assembly Bill 2098, a law that would revoke the licenses of any physician who utters opinions "contrary to the established scientific consensus."[1] If that's not Orwellian, I don't know what is. Luckily, discernment still occasionally exists in the world and a judge shut it down.

My Covid expertise, borne of obsessive deep study, led to the uncovering of numerous scientific frauds around the suppression of ivermectin and hydroxychloroquine, and the introduction of fraudulent medicines like molnupiravir, remdesivir, and Paxlovid. Being forced to observe those frauds become embedded into public and health policy around the world and see millions suffering and dying as a result was not only frightening and dystopian, but also motivating.

As if the suppression of effective early Covid treatments was not enough of a catastrophe, I watched as an even bigger disaster unfolded due to the toxicity and lethality of the global vaccination campaign and the shocking, sudden ignoring of natural immunity. I saw masses of people all over the globe being fed unrelenting propaganda, whipped up into such fervor that they not only became complicit in their own (sometimes fatal) injuries, but also spurred into demonizing, ostracizing, and destroying the livelihoods of people who refused the vaccine.

In what would unquestionably become a high point, in February of 2022, three months after my ICU career ended, I was being deluged with questions and requests for treatment from both the vaccine injured as well as those suffering Covid long haul syndromes. Paul and I had been immersed in studying the mechanisms and therapies for treating these spike protein-induced diseases; in fact, Paul's I-RECOVER post-vaccine treatment guide is the most comprehensive of its kind, with 425 scientific references at this point.[2] I knew that there were hundreds of thousands of people suffering post-jab and that the system was either ignoring or dismissing them; similarly, Covid long-haulers were simply being shuffled from referral to referral and rarely receiving any sort of treatment or relief. I decided to change my specialty to the study and care of these two groups and opened a specialty tele-health practice, an endeavor that has been both incredibly challenging and uniquely rewarding.

Kristina Morros of the FLCCC took on a second job as my private practice manager, and I am eternally indebted to her. Kristina literally built the whole practice with me from scratch. Another gift from the universe was a nurse practitioner extraordinaire named Scott Marsland who we hired to see patients as he had also developed an interest in treating the vaccine injured and Covid long-haulers. Our initial electronic health record system invited all of our patients to submit reviews after the first visit, and I had never read such glowing patient reviews about any practitioner in my life. I was actually jealous. Although my patient reviews were and are strong, they pale in comparison to Scott's. I knew right away that he was a keeper and would be immensely valuable both to me and our patients.

By the end of the year, Scott and I had formed not only a close clinical collaboration sharing insights, successes, and failures, but we went on to form a formal partnership and created a new corporation together. We also lucked up and hired the most caring and courageous nursing staff led by Tisha Palmer, Kara Gabrielson, and Michelle Kuklinski, all of whom had lost their jobs because of their refusal to get jabbed. We later added two additional accomplished nurse practitioners, Heidi Dumas and India Scott, as well as another nurse, Karen Gates. Kristina ended up going full-time with the FLCCC and now Tisha is killing it as my practice manager and lead nurse, and our staff continues to grow.

My practice has been filled with hundreds of formerly healthy, success-ful people in the primes of their lives who were not only now disabled, but severely so. I cannot stop thinking of the scores who died before they could even make an appointment.

I have observed tremendous anxiety and depression in the injured, not only over what had become of their lives and their health, but also over their futures. Imagine feeling as if you were dying and not knowing if you would ever get well, all while being abandoned, patronized, or gaslit by your medical providers.

As the world slipped ever deeper into propaganda-fueled madness, my patients would relay endless stories of system providers devoid of empathy in their encounters. As if I wasn't already estranged from medicine and my fel-low physicians, the void between us became even greater. I could no longer recognize them as anything but arrogant monsters, ignoramuses, cowards, or all three. They were and are the antithesis of my mission, which is to protect and treat as many people as I can.

For a solid year and half, the words "vaccine injury" could not be uttered by any system provider lest they swiftly be branded with a scarlet A for "anti-vaxxer." (I hear from sources on the inside that this is some-what less taboo and even discussed a bit more openly now; I pray this is both widespread and true.) This revolting slur was attached to vaccinated people who opposed mandates as well as those who had gotten their jabs and been injured and were simply trying to warn others of potential risks. *Anti-vaxxers.* Yup.

With the creation of my practice, I proceeded to smash my former records of work burdens and busyness, even those set earlier in Covid. I knew I was living in a truly historic time, and I believed that I could make a dif-ference. My work and advocacy have turned me into a public figure, which complicates things for a chronic yes-man. With millions dying preventable deaths . . . *I. Could. Not. Say. No. To. Anything.* Lives literally depended on me saying yes. I wasn't looking for fame or celebrity; those have never been my goals. I do not get dopamine rushes from speaking or being promoted publicly, if anything, the appearances and interviews exhaust me. I do enjoy them in the moment, given the conversations are always interesting, and meeting courageous, intelligent, and profoundly studied people is morally and intellectually rewarding. As a life-long educator who enjoys teaching and sharing knowledge, never had I encountered a classroom so immense or a curriculum so critically important—especially the data we were uncovering around the vaccines.

Relative to many of my peers, I realize that my Covid lows could have been a hell of a lot worse. And fortunately, the highs were many and impactful.

By the end of 2021, the impact of the FLCCC across the world was truly astonishing. We had eight million protocol downloads in twenty-four languages, 33 million website page views, 100,000 newsletter subscribers, 600,000 social media followers, four million webinar views, and over 13,000 donors. We were on fire.

I started my "Medical Musings" blog on Substack,[3] where I began to write on numerous aspects of both the ivermectin and vaccine frauds. It was part journal, part expose, and I didn't know it at the time, but it was building the foundation for this book.

Kelly Bumann took over as our Executive Director and then later Zahra Sethna as Senior Director of Communications. Together they took expert control of the mothership, building out a team of cybersecurity, website design, and social media experts as well as hiring Kate Vengrove, our talented Director of Development and an absolute fund-raising force.

My social and professional network of researchers, clinicians, writers, podcasters, health freedom advocates, and journalists absolutely exploded. I was meeting and collaborating with incredible people from all over the world, people like Brian Brase and Ron Coleman of the People's Convoy in the US; Dr. Maria Hubmer-Mogg in Austria; Sheri St. Louis, Byram Bridle and Paul Alexander of Canada's Freedom Convoy; Dr. Jennifer Hibberd, Ira Bernstein, and David Ross from the Canadian Covid Care Alliance; Didier Raoult and Louis Fouche in Marseilles, France; Tom Borody in Australia; Andrea Stromezzi and his colleagues in Italy; Sture Blomberg and his colleagues in Lakaruppropet from Sweden; and of course, Medicos Pela Vida in South America. All have been attacked and persecuted for early treatment advocacy.

A quick note on Jennifer Hibberd and the Canadian Covid Care Alliance (CCCA), which I regard as a sister organization to the FLCCC in terms of mission and spirit. Jennifer, a dental surgeon and academic educator, came across Dr. Rajter's study in *Chest* in October 2020 which showed the astonishing reduction in death among ivermectin treated patients. She shared the study with her brother, a consultant to Newsmax, who brought it to them to disseminate its findings more widely.

Around the same time, Jennifer heard of the nursing home in Toronto that had been using ivermectin against scabies before the pandemic began, and whose residents fared extremely well when Covid hit. She began joining Zoom calls with doctors in South America and elsewhere who were reporting the lifesaving results they were seeing with ivermectin. She began speaking publicly about ivermectin on social media and was subsequently drawn into the Ivermectin MD group on Facebook where pioneers of the medication's use were brainstorming ways to help Covid patients across the world. When the Facebook group inevitably got shut down, Jennifer turned her attention to launching the CCCA with Dr. Ira Bernstein and David Ross. Later, Jennifer and Tess launched the World Council for Health, which among other things provides expert guidance on preventing and managing Covid safely at home.[4]

To my knowledge, Jennifer was the first Canadian to recognize and promote ivermectin for Covid. It's impossible for me to positively identify the first American to do so, but two individuals come to mind. One is Eleanor Watson, who posted a strong mechanistic argument on Facebook in early February 2020.[5] The second is a Harvard trained lawyer I met later in my journey named Jay Sanchez. Raised in Bolivia, Jay was already familiar with ivermectin and its uses for various ailments. After studying several therapeutics, he decided the best candidate for Covid was ivermectin based on its success with similar types of viruses. In late March of 2020, he drafted a four-page argument supporting its efficacy and sent it to some virologist friends at Harvard. Despite his heroic efforts—which included putting flyers on car windshields in parking lots—he did not gain any meaningful traction beyond a few of his friends going out and buying themselves a stash of ivermectin.

The Defeat the Mandates rallies in Washington, DC and Los Angeles were an undeniable high. The day after the DC rally, Senator Johnson held a roundtable called "Covid-19: A Second Opinion," where a group of doctors, scientists and medical experts were invited to speak on early treatment, vaccine safety and efficacy, and basically what the world had gotten right and wrong in Covid. It was a five-hour meeting, and it allowed me and the rest of the most public and expert Dissidents to perform a full take-down of the two years of insane non-scientific policies the country had been subjected to. Ron told me that 90,000 people watched it live, for the full five hours, with nary any drop off even near the end. As of this writing, a shortened thirty-eight-minute video on Senator Johnson's (reinstated) YouTube channel has 2.6 million views.[6] Paul's tearful speech describing what his hospital did to him and his patients is a must-see.

Another high point for FLCCC was when, in March 2022, Paul received a unanimous commendation by the Virginia House of Delegates for "his courageous treatment of critically ill Covid-19 patients, and his philanthropic efforts to share his effective treatment protocols with physicians around the world."

The FLCCC was also doing lots of work under the radar. We were contacted by Dr. Thomas Schirrmacher, the Secretary General of the World Evangelical Alliance which connects Protestant churches belonging to 143 National Evangelical Alliances with a total of 600 million members. Only the Catholic Church is larger, thus we called him "the second Pope." He had heard of our group and our work and sent our papers to a trusted German academic friend of his, who reported back to him that we and ivermectin "were legit" (my words). Dr. Schirrmacher then assigned several colleagues, on a miniscule budget, to try to get our protocols to protestant organizations across the world. Although his efforts were heroic, they ultimately made little impact, as they were no match for the globally weaponized health agencies and media propagandists.

Another notable high point was when the FLCCC's Keith Berkowitz, one of our founders and a board member, was approached by a patient who had prominent contacts in some Palestinian health agencies. Keith's patient proposed to us that we develop a simplified treatment protocol for mass distribution in Palestine (the Palestinians were very reluctant to take the Israeli vaccines), which he would then personally purchase for mass distribution there. We worked hard on that effort, developing a pared-down protocol called I-RESCUE. The plans for that initiative were scrapped when the donor abruptly decided against pursuing it in the wake of the FDA's horse-dewormer tweet.

Similarly, I gave highly visible lectures in Indonesia for a woman who owned a factory that manufactured ivermectin. Sincere and courageous, she was trying to get the drug approved and deployed widely as Indonesia was getting hit hard. She knew ivermectin was effective as many in her orbit had been saved with treatment. Tragically, after the lectures and press conferences I did there, her factory was raided, and she was left immersed in political and professional turmoil.

However, through another connection, Paul and I were invited along with several other physicians on a private Zoom call with Sri Lankan President Gotabaya Rajapaksa. He told us he was impressed with our message, guidance, and expertise and said he would do what he could to influence his public health officials. Not too long after, Sri Lanka approved ivermectin for off-label use. It was a huge win for the "Pearl of the Indian Ocean" and its people.

I went to DC several times, speaking to the House Freedom Caucus, Ginni Thomas's Third Century Group, and also the Congressional "Docs Caucus." The latter was on the invitation of Florida Congressman Dr. Neal Dunn, who had contacted the FLCCC to thank us for our protocols. He also reported that he had personally treated over 200 congress members, staffers, and family members with our protocol, and all had done exceptionally well. All were Republican. He did not want me to make his name public at the time, so I just tweeted that over 200 members of Congress had been treated with ivermectin. Boy, did that cause a storm on social media. Some saw it as a "let them eat cake" moment, as if politicians were treating themselves with ivermectin while denying the public access to it. Nothing could have been further from the truth—but try explaining that to an irate Twitter mob while protecting your anonymous source. My favorite comment, by a nobody with 645 followers and more than 30,000 tweets, was this predictable bit of drivel: "I'm sure this is in an RCT, double blind study that's peer reviewed and published, right?"[7] A few journalists even reached out, but because I could not reveal my source, they did not report on it.

Then an agent from one of the intelligence agencies contacted me. After following the FLCCC's and my work for a while, he smelled corruption at the federal health agency level and wanted to open an investigation. He was intelligent and insightful and upset by what was going on. A young father and early in his career, he clearly wanted to make a difference, but I knew that as soon as he ran his proposed investigation up the proverbial flagpole, it would be dropped faster than a flaming potato. When he ghosted me with no explanation after several interviews, I had to assume I'd been right.

We held international press conferences, moderated by widely known medical educator Dr. Mobeen Syed ("Dr. Been"), with a panel of Covid experts giving short summaries of our expertise on ivermectin and early treatments. We corralled as much global press attention as we could but once again, it was the Bad News Bears against the Yankees. A couple of ripples went out, but it was never the tsunami we were hoping for.

I was invited by a journalist from Mexico named Gabriela Sotomayor to give a private presentation to the United Nations Press Correspondents. One of the most prominent international journalists helping to spread the word about ivermectin, Gabriela wrote several supportive articles profiling both me and Juan Chamie in a widely read weekly newsmagazine in

Mexico called *Proceso*. Although Gabriela worked valiantly to drag the UN correspondents to water, she could not make them drink. I received not an iota of support—not even a request for more information—from the UN.

Some events and experiences represented both ends of the high/low spectrum. Paul was invited to lecture to a physician organization in India with an audience of over 20,000 doctors on Zoom. His presentation was extremely well-received. (High.) Contrast that experience to when he gave the identical lecture to the Department of Anesthesiology and Critical Care at Harvard University. The organizer of the latter let Paul know afterward that he had been deluged with complaints that Paul's lecture was full of misinformation. (Low.)

We learned that our MATH+ hospital protocol had become the official treatment protocol in the Ukraine,[8] and I traveled to Puerto Rico to participate in the drafting of the fourth declaration of the Global Covid Summit.[9] This declaration eventually went around the world and obtained 17,000 signatures from physicians and researchers (all extensively vetted/verified). The press only touched it to declare it illegitimate.

My lectures and our protocols in Malaysia caught the attention of a prominent Malaysian businessman, Tan Sri Lee Kim Yew. He personally purchased large quantities of ivermectin to distribute in his country and recognized the FLCCC with an award for our efforts and advocacy in Covid. His organization, the Cheng Ho Multicultural Education Trust, presented us with their Benevolent Award in 2021. The Cheng Ho award came with a $20,000 grant and a future promise of $100,000 to fund a collaborative public health project in Malaysia. A beautiful vase and painting accompanied this generous and humbling award. I also collaborated with several nonprofit organizations that supported the indigenous peoples of the Amazon, giving lectures presenting the evidence of ivermectin's efficacy as the groups were trying to raise funds to purchase the therapeutic in large quantities for distribution.

I participated in a moving and impactful vaccine injury conference in Ohio, sponsored by the state's CHD chapter. Paul's speech was absolutely memorable again.[10] We were also invited by Bobby Kennedy to speak at CHD's annual conference this year, an incredible event and an even more gratifying honor.

Then came the amazing documentaries I was asked to participate in. I have bottomless respect for the truth-telling filmmakers who were working to expose the realities of the gruesome situation we all found ourselves in, and I would put *The Real Anthony Fauci, Plandemic 3,* and *The Truth About Ivermectin* by Mikki Willis at the top of my must-watch list.

Most recently, I was invited with Peter McCullough, John Leake, and the incredibly courageous Australian doctor Melissa McCann on a speaking tour across Australia, an effort funded by one of the more amazing people I have gotten to meet in my Covid journey, Australian billionaire Clive Palmer. I could write an entire book about Clive and someday I might; in the meantime, I'll have to summarize his determination and generosity the best I can: In early 2020, with the permission of the Australian government, Clive bought millions of dollars' worth of hydroxychloroquine to distribute to the Australian people. By the time his plan had made it to the execution stage, the War on Hydroxychloroquine had begun. In response to a tsunami of fraudulent trials, the Australian government destroyed the entire supply of HCQ.

In the US I was interviewed by Eric Bolling on Newsmax and Maria Bartiromo and Laura Ingraham on Fox. Internationally, I did TV interviews and podcasts in Norway, Australia, UK, Ireland, India, Hungary, Australia, Italy, and Canada (to name just a few). I also spoke at Covid conferences in Brazil, Austria, Italy, and France—including a speech I gave in my rusty French to a large outdoor crowd at the second annual International Covid Conference in Paris.[11]

In October of 2022, we put together the FLCCC's first medical conference on treating Covid-19 vaccine injury and long-haul syndromes called "Understanding and Treating Spike Protein Induced Diseases." It was a huge success, drawing over 325 attendees, the majority physicians. Many shared tales of persecution or suspended licenses or both and spoke of having to reinvent their careers in the wake of their livelihoods having been destroyed. The connections, collaborations, and sense of community we shared will stay with me forever.

One by one, we Dissidents started getting asked to testify or support bills drafted by various courageous state legislators and health freedom groups in Tennessee, Virginia, New Hampshire, Maryland, and Ohio. These were all bills specifically created to combat or protect citizens from the Feds' onerous policies. Probably our biggest win was getting Tennessee to pass a bill making ivermectin available over the counter. Amazing.

Richard Fleming and Robert Malone recently testified in Texas for a handful of bills intended to protect physicians and their use of off-label medicines without penalty. And just three weeks ago, I testified in support of blocking the mandating of both the meningococcal vaccine and the chicken pox vaccine as part of Wisconsin's childhood schedule. The motion to block the mandate passed.

In Clown World news, Paul Marik, Mary Talley Bowden, and Robert Apter filed a lawsuit going after the FDA for their horse tweet and for their illegitimate attempts at influencing the practice of medicine in regard to ivermectin. Led by Clayland Boyden Gray of the prominent DC law firm Boyden Gray & Associates, the suit compellingly pointed out that the FDA has no authority to interfere with the practice of medicine. The case was (wrongly) dismissed after a liberal judge ruled that the government can't be sued, but we are very hopeful on appeal.

As of March 2023, the FLCCC had updated our protocols almost a dozen times to evolve with the literature emanating from over 800 peer-reviewed publications. We have created and published new protocols on treating RSV, flu, metabolic syndrome, and diabetes. Since inviting "Dr. Been" to join our core team last year, he has produced more than sixty episodes of his brilliant "Long Story Short" series. Paul is now working on adjunctive treatment protocols with repurposed drugs and nutraceuticals for cancer—because if there's a giant bear anywhere in the vicinity, you can count on Paul to poke it. I love that about Paul. We also plan on creating an extensive educational campaign and protocol for sepsis centering around the use of IV vitamin C. I might even start writing another book called, *The War on IV Vitamin C*, if Jenna is up for it. [Coauthor's note: *Jenna is very much up for it.*]

In early 2022, we had over nine million page views on the FLCCC website from 230 countries. Media impact, despite the relentless censors, has been formidable as well. From April 2020 to March 2023, Christensen Public Relations estimates that our "earned media" mentions (not counting YouTube, Rumble, Odyssey and the like), reached an audience of 965 million, from 726 interviews in fifty-seven countries. Whoa. My Fox News op-eds alone reached more than six million people.

Probably one of the most meaningful moments in the FLCCC's history was when out of the blue, we received an incredibly gracious letter from the Nobel Prize–winning discoverer of ivermectin, Professor Satoshi Omura, and the Kitasato Institute. He also gifted us with a stunning spectrograph of the species of bacteria from which ivermectin is derived. I framed both the spectrograph and letter, and they hang over me as I work from my desk. And they always will.

The letter reads:

Dear Dr. Kory

I heard about your group's activities from Dr. Yagisawa. The barriers that stand in our way are high, with businesspeople and regulators who place priority on their own interests, but I am confident that your work, supported by your dedication to humanity and science, will overcome these obstacles in time. At the Kitasato Institute, we are working to have Ivermectin approved for the treatment of COVID-19 in Japan as soon as possible using the information that you provided.

I have enclosed a collotype photo of Streptomyces Avermectinius, the only microorganism that produces Avermectin, as a token of gratitude for your work. I hope that this photo will watch over your work.

I wish you good health and success of the FLCCC Alliance activities.

Sincerely yours,
Satoshi Omura, PhD

It is my belief that the war on ivermectin is all but over—or at the very least, at a stalemate. The doctors across the world who are using it to treat COVID will continue—unless we lose our medical licenses to do so, like they are trying to do to my friend Meryl Nass, a physician from Maine who has been criminalized for promoting and prescribing HCQ and ivermectin. (Paul Marik and I are both serving as her expert witnesses in the hopes of reversing the suspension.) The rest of the system docs will continue to dismiss ivermectin as a horse dewormer and prescribe Paxlovid instead. It's been a brutal, bloody war, and although we cannot claim victory, our message and guidance reached and saved millions of lives. We are pretty damned proud of that.

It's hard not to wax philosophical when I look back on my crazy, complicated, unexpected Covid journey thus far. I've held the hands of patients and families suffering in the depths of despair and celebrated with others at the height of jubilation. The people I've met, the camaraderie I've experienced, and the skill and sacrifice I've seen all around me are profound and humbling. I genuinely don't know what's in store for me or the FLCCC next (if the last three years have taught me anything, it's that the universe has a mind of its own), but I'm excited to find out.

CHAPTER FORTY-THREE

Testimonials

The evolution of evidence-based medicine into what I now call "evidence-based mania" is a result of what I see as a near complete disavowal of the methods we've relied upon to determine new therapies for hundreds of years. These include a combination of knowledge of pathophysiology and pharmacological mechanisms, empirical data, and the clinical experiences gleaned from close observation of hundreds if not thousands of patients with a similar disease or symptoms.

When I read books about the history of medicine or medical papers published fifty or one hundred years ago, I am regularly shocked by how much they knew despite having access to a fraction of the scientific, diagnostic, imaging, and clinical trial methods we have today. They did not rely on randomized controlled trials to discover and deploy new therapies; they relied on observation and experience. Did they get everything right? Absolutely not—but neither do we today. Still, I am convinced that if this pandemic had occurred a hundred years ago and ivermectin had been in use, knowledge of its efficacy would have spread like wildfire, simply based on the clinical experiences of the bedside doctors. Now, no matter how many hundreds or thousands of doctors report efficacy, it gets ignored until an official RCT then "proves" them wrong. The big RCTs are so powerful, they could convince an elephant he was a hyena.

The testimonials below bring actual tears to my eyes; not only for the deep gratitude expressed and compelling outcomes described, but also for all those who were denied the opportunity to be treated by a physician willing to use one of history's safest medications in a deadly disease.

A quick disclaimer (or two): In a number of these clinical anecdotes, people describe sneaking in ivermectin to a hospitalized family member, sometimes using veterinarian formulations. Politically and legally, I cannot endorse such actions; however, I admire and commend those who refused to sit by and comply, and I feel it's important to document historically the lengths that loved ones had to go to in the face of widespread and dangerously restrictive hospital and governmental policies. Additionally, the FLCCC (and I) get emails and messages from around the world, so oftentimes, English is not the author's native language. Therefore, I have taken the liberty of cleaning up only grammar, spelling, and punctuation issues within the following messages. I would never dream of altering them in any other way; they are too perfect and powerful the way they were written. I hope you find them as moving as I do.

This is from the first patient I interacted with who had been treated with ivermectin as a result of learning about my preprint review paper. She later wrote to Senator Johnson to have her case history included in the Congressional Record along with my testimony delivered on December 8, 2020.

Hi Senator Johnson,

I developed COVID in early November and was sick for two weeks. I ended up getting COVID pneumonia and had a resting heart rate of 125-135 and chest pain. I had a second trip to the ER and a chest CT diagnosed pneumonia. I had a fever on and off for two weeks and ended up with a fever of 102.7 on my second ER trip. I was sent home and told to come back if I had increasing shortness of breath and fever.

I researched COVID protocols online and found the I-MASK protocol at Eastern Virginia Medical School. I ended up speaking with Dr. Kory and sending the protocol to my pulmonologist who prescribed Ivermectin. I had a fever of over 102 the night I took the Ivermectin and a resting heart rate of over 120. I took the ivermectin medication at 6 p.m. By 10 a.m. the next day, my temperature had returned to normal and so did my heart rate.

My 92-year-old mother ended up getting COVID after being exposed to me prior to me being aware I was sick. She has a pleural catheter and is terminally ill with advanced heart and lung disease. We were working to place her in hospice. Dr. Kory prescribed Ivermectin. I went over with a pulse ox and her oxygen level was around 85. She was extremely fatigued and short of breath but did not want to go to the hospital. She told me if she died, she wanted to die at home. Within two days of taking two doses of Ivermectin, her had O2 returned to normal which is simply amazing. She is terminally ill but will not be dying alone in the hospital with COVID. We are starting hospice next week. Neither my mother nor I experienced any side effects from the Ivermectin.

Warm regards and well wishes.

This email is the first of two from the same man:

Dear Dr. Marik,

My father is still requiring significant oxygen. They have him on prednisone. I will continue to try and sneak in the ivermectin. That's all I can do. They didn't even bother to look up the work you are doing or all of your publications, and even went so far as to say you might be a store clerk??? HORRIBLE! All this after I printed out all of the data and studies and your protocol and handed it to them days ago!! Needless to say, I am distraught. I said that doctors all over the country are adopting your protocol, why not Naticoke Hospital? Again, they said the clinical trials are not there and they won't budge. Dr. Marik, this is a crime against humanity. I love you for the work you are doing. I pray my dad lives. I don't know if I can pursue legal action. I am sick about this.

Two days later, he sent this:

Hi Dr. Marik!

I have an update for you! So, on Tuesday [my father] was on a high flow cannula of 25 L of oxygen. I gave him a full dosage of ivermectin on Tuesday and on Wednesday his oxygen went down to 10 L of low flow oxygen. Yesterday, I gave him another full dosage of ivermectin in his strawberry milkshake and today his oxygen needs are down to 2 liters!!!!!! The pulmonologist, Dr. XXX is very unprofessional. She STILL hasn't returned my calls. In fact, she never spoke to me over the two and a half weeks that he has been there!! She missed the opportunity to learn from you! I'm just so pleased that my father is on the mend. My father will be going into a rehabilitation center after this so I'm going to do my best to educate them in order to get him the best care. I heard from a nursing friend that New York is using ivermectin! Yay!! I would be happy to give a testimony and I will be telling everyone I know my story. Bless you and thank you so much for everything you are doing! Thank you for responding to my email and calling me! I am saying this with tears in my eyes. I was trying to get ivermectin in the very beginning of his hospital stay, but it was so hard. So grateful that it worked even on day 14 of his hospital stay! Thank you, thank you, thank you!

This message summarizes the fear many expressed to us in even considering a treatment the media had made out to be not only ineffective but potentially dangerous:

Dear Dr. Kory,

In July, my whole family and I got Covid. My dad is 72 with atrial fibrillation (irregular heartbeat) and hypertension and mom is 70 years old. Two of us are in our 40s.

The Covid symptoms presented rapidly overnight for all four of us. My dad tested positive just before the weekend and the rest of us over the weekend. Due to this we had to wait for Monday to get IVM at a compounding pharmacy. The three of us started on Monday.

For my dad we were worried as he is on blood thinners, so he initially didn't take it. I was so scared that I decided to email Dr. Kory. Within 6 hours he replied. I nearly fell off my chair as honestly, I didn't think I'd get a response. His opinion was that the best treatment for Covid is ivermectin.

We took my dad to a doctor in South Africa the next day, one of the few prescribing practitioners, and on day five he started treatment. We had to leave our GP of over 30 years to get treatment from another doctor. He monitored my dad's INR [internal normalized ratio, a measure of blood clotting speed] and d-dimer levels to make sure his blood wasn't too thin. My dad didn't have to stop his current blood thinners.

As my dad was positive before the rest of us, he was already further in and we could see he was taking so much strain from the effects of the virus. I swear if we didn't get him the ivermectin when we did, he would have had to go to the hospital. I think we caught it just in time as overnight you could see the difference in him and could hear it too. He could feel it also. He cleared up better than all of us. He took five days of treatment at a higher dose and was back at the office two weeks later. The rest of us only took three days of treatment and we all swear by ivermectin and think it is the number one treatment for Covid.

You won't believe the fear I had even putting that initial tablet into my mouth. The way this medication has been portrayed and the fear mongering around it actually causes undue stress at a time when people shouldn't be stressed questioning a safe, proven treatment.

I thank Dr. Kory for his time and his reply. You all deserve the Nobel peace prize [sic] and the work you are doing helps many people. I was lucky to see a recording of Dr. Kory speaking last December, and if I hadn't, I wouldn't have known about you all and the protocol. I have since passed the FLCCC details onto everyone who is willing to discuss Covid options.

Thanks Dr. Kory . . . you are an amazing doctor. Your kindness to email me back, I truly feel was a moment in my life where if you didn't hit that send button, my life and my dad's could have been very different.

You all rock, IVM rocks, and the good Doctors left in this world rock. Your morals and ethics to first do no harm . . . and courage to stand up and fight against the system inspires me daily.

Keep shining. Stay safe. Kind Regards.

This was an email from an acquaintance:

In the mood for a powerful testimonial of ivermectin?

The wife of our president has had Covid for going on three weeks. She has had a fever over 102 for two weeks. I can't describe all the details of what she shared with me as far as how badly she hurt and the pain she was feeling. She thought she was dying.

Monday morning, she was disoriented, feverish and felt like vomiting and went to the hospital.

Her white blood cell count was so low they said they were going to have to isolate her but there was nothing they could do for her. She refused to stay at the hospital, so they sent her home under the condition that she committed to isolate herself.

She even asked them for Ivermectin, because of former conversations with me, and the doctor refused and told her not to believe the lies on the Internet.

They sent her home with nothing and said, if you get to the point you can't breathe, come back and we will put you on a ventilator. (She has enough of a medical background that she knew that was a death sentence.)

She said she had been so weak she couldn't even stand up and take a shower without passing out. The pain, as she described it, was almost more than she could bear for the past two weeks. She had almost resigned herself to the fact that she was going to die.

Her husband called me Monday morning after dropping her off at the hospital and hearing their prognosis.

I drove to their house at lunchtime and delivered Ivermectin and all the vitamins to their house. The husband took his first dose. By the time she got home from the hospital her husband had her dose waiting for her.

Within 48 hours the fever was entirely gone. She says the contrast of how quickly she felt better was dramatic. She broke down in tears bawling and thanking me and telling me to thank you for being so generous to share this knowledge.

This email came from a colleague within the FLCCC:

Hi Pierre,

I have an incredible story. I was contacted urgently Thursday evening by a 44-year-old with COVID-19 on DAY 12 at home, extremely dyspneic [short of breath]. He had been to the ER three days prior. He asked about ivermectin and was told by the doctor that it's for horses. They threw him out and said he'll be back in three days on a ventilator. They were mad because he was unvaccinated. When I tele-assessed him, he could only speak in short sentences. I knew this was serious.

A friend had given him ivermectin, but he hadn't taken it. I told him to take it immediately. I ordered home oxygen, and the company was at his home within the hour. I ordered more IVM from the compounding pharmacy and started him on prednisone, spironolactone, and dutasteride as per the new protocol. Within hours he was feeling better.

Yesterday morning his girlfriend wrote to me:

Good morning Dr.

I have some heartwarming news. He said that overnight he has felt relief already. He feels like whatever the virus was trying to do to his body is starting to finally reverse course. You are an angel doing God's work.

I just got off the phone with him. He sounds fantastic. I have never heard such a grateful patient. He says he will do anything to help us. As the evil at the FDA, CDC, NIH, WHO and Health Canada permeates, we just have to remember the good work that we do! Take care.

A man called into the FLCCC to get ivermectin for himself, and wound up sharing the following story, transcribed by an FLCCC staffer:

This man's father was admitted to the University of Tennessee hospital critically ill with Covid. They gave him one dose of Ivermectin—ostensibly to see if his symptoms were being caused by any parasitic diseases. When he did not improve, his family was told he might have to be placed on a ventilator. The physician informed the family that due to extensive lung scarring, the dad would not likely survive. They called in hospice and began to prepare the family for his eventual death.

In the meantime, because his dad was not yet on a ventilator, the son procured some ivermectin from a friend who had ordered it from India. He gave it to his father on Tuesday and Wednesday —and of course, did not inform the hospital. There was no immediate improvement. But this [Friday] morning, the father texted the family from his hospital bed to tell them that he had made a significant turnaround. His doctors told him they had never seen anything like it and didn't understand what caused him to turn around so quickly and so dramatically. They told him he would be discharged by the end of the week or the weekend because of his "miracle" turnaround.

* * *

Drs. Kory and Marik,

I am a 49-year-old, type 2 diabetic (diabetes diagnosed after being admitted to the hospital for Covid-19) and was hospitalized for 7 days with Covid-19 in September 2020. Two weeks ago, I was diagnosed with Covid-19 with watery eyes, headache, lethargy, runny nose, backache, and nasal congestion.

I was determined to find a treatment as I did not want to end up in the hospital again. Researching online, I was surprised to learn that there were very few therapeutic advancements to treat Covid-19 until I came across your protocols using ivermectin.

I found a nurse on your website who introduced me to Dr. XXX, a family care physician in Texas who follows your Covid-19 protocol. Twelve hours after taking the ivermectin, along with the multi-vitamin, vitamin D, antiseptic mouthwash, and melatonin, my symptoms improved. Twenty-four hours later my symptoms significantly improved, and after 36 hours, I felt 100 percent normal! I wanted to thank both of you for following the science and generously sharing your protocols.

I've been sharing my experience and referring patients to Dr. XXX. I've reached out to my close friend who is the lead investigative reporter for CBS in Dallas, and she's interested in possibly doing a story on ivermectin. Would you be interested in visiting with her?

* * *

Thank you for your response to my email. My husband (a retired allergist) prescribed Ivermectin for our daughter since neither her PCP nor her oncologist would do it. Her first dose was Tuesday night. After two weeks of having a fever, migraine-like headaches and insomnia, her fever was gone in the morning. After her second dose, the headaches were gone, and that night she slept for almost ten hours. She regrets not taking it sooner.

Thank God and for you and the other doctors who are enlightening those of us who search for the truth.

* * *

You saved another life. A friend of a friend was circling the drain till she was able to get some farm grade ivermectin paste. No matter what happens, be assured YOU are on the VIP list for heaven.

* * *

Pierre,

I was just reading the Ivermectin Trial Site News. We read about the Fype and Bucko families. Thank you for caring and fighting. It is unbelievable when doctors refuse to help a patient in need. I hate to say that we saw that when our son was sick. Thank you for caring and fighting for people.

We helped a family last week whose father was in the hospital with Covid, pneumonia and blood clots in his lungs. We helped them get Ivermectin to him. He was out of the hospital in two days and is doing much better.

Thank you for your work. We know it is not easy.

* * *

Merry Christmas Dear Doctor.

I am the only one in my country who uses this protocol. I am the only member of FLCCC from former Soviet Union countries. I am very proud that I started using this protocol and I can see faster recovery from Covid-19. So far, I have used it in many patients. It helps. Unfortunately, we do not have Ivermectin in our country, but I would like to say that protocol works very well even without Ivermectin. I am a cardiac anesthesiologist and intensivist which allows me to work on a different level with critical patients. I would be proud to have my name in your paper and will be very grateful to be in touch with very knowledgeable people like you.

With warmest regards.

* * *

Dr. Pierre Kory,

On December 8th, 2020, I had the pleasure of watching your testimony in front of the senate panel regarding your research on Ivermectin and how it treats Covid-19.

I somehow was able to reach out via email and you responded that evening to me by phone. My mother-in-law, a school nurse, was sick with Covid-19 and was not doing well with it as she is older with some health issues. You were able to call her in a prescription for Ivermectin and it was truly amazing how fast it worked for her! This medication made not only us but many others true believers in your research for Ivermectin.

Ivermectin should be used to treat Covid-19, as we know that it does work!

* * *

I live in GA. I am a nurse at a very busy high school. So, with that being said, I contracted Covid 19. I became very sick, and it was very fast moving. The doctor gave me antibiotics to try to prevent pneumonia. I had fevers, shortness of breath and was very weak. My oxygen was low, and we were scared I might have to go to the hospital.

My son-on-law saw Dr. Kory addressing the senate about Ivermectin. He was so amazed that this medication could help me. He called the doctor and left him a message. The doctor called him back and explained how the meds work. Mike told him about me being sick and that I am a nurse. The doctor called the Ivermectin into my pharmacy. By the next day I was so much better. The fever broke and I was able to get out of bed. I was shocked at how much better I felt just overnight. I took the second dose 48 hours later and again felt better. I went back to work after 10 days instead of 14 days of quarantine. I know the Ivermectin helped heal me from this virus.

My boss said that I recovered twice as fast as the other employees who had gotten sick. I later walked into her office, and she was telling other people about this medication and how fast I recovered.

Anyone reading this please keep an open mind that it does help, and many people have used it. They should be able to write the medication for everyone.

I hope this helps if you are undecided about the medication.

* * *

Pierre Kory, you are amazing! Firsthand testimonial on ivermectin: We took it and it kept my parents from getting really sick. All of us had it; I and two of my kids have asthma. I couldn't breathe. Covid is the [worst] virus I have ever had. Dr. Kory was able to get ivermectin prescribed for us [and] within 24 hours we could feel a release on the breathing. A friend of my dad's who infected us passed away last weekend from Covid. Ivermectin kept us safe and killed the virus.

* * *

I got my medicine at the only pharmacy I'm aware of in Illinois that will fill this prescription. This pharmacy is listed on your FLCCC website.

I would love to tell you what happened within 24 hours on the tenth day of being very sick, if only to thank you. I want you to know, patriots all over the world thank you. If you're ever in Chicago I will buy you a drink.

I listen to every interview that I can find that you have given. I want you to know, not all heroes wear capes. Some wear scrubs and white coats. I want you to know, you are a HERO. God bless your work. You are amazing. Thank you from the bottom of my heart.

* * *

Good evening, Dr. Kory,

I watched your interview before congress and on your website. I went ahead and bought [ivermectin] in Tijuana, Mexico and just started taking it. I called my friend who I had not seen for a few weeks. She told me that her husband who works at a hospital got Covid and was sent home to only take Tylenol. She said that he was very congested and had a bad cough. I gave her Ivermectin along with the protocol and she just told me that her husband is recuperating incredibly well.

Thanks so much for being at the front of it all and providing us with information on how to fight this virus.

* * *

Dear Dr. Marik & the FLCCC Team,

I would like to take the opportunity to thank you all for your great service to humanity during this Covid pandemic. Your selfless hard work is saving lives. A dear friend of ours responded almost immediately to the Ivermectin protocol that the FLCCC put out very recently. He was in a very bad way and basically had all the symptoms of Covid and was at the point of giving up hope. The Ivermectin that he took was for Veterinary use and it worked. Doses were calculated correctly by his veterinary doctor (not a joke). The human version of Ivermectin is not available is South Africa and in fact has been banned from being imported by the South African Health Products Regulatory Authority (SAHPRA). Crazy times.

I have known about the FLCCC Team since the early days of the pandemic, as I followed Dr. Chris Martenson of Peak Prosperity as he tracked the FLCCC. He did an amazing job reporting on the scientific data coming in on the virus on a daily basis. I was made aware of the MATH and MATH+ Protocols via him. I did watch Dr. Kory's Testimony to the Homeland Security Committee and realized through his conviction that Ivermectin had real value in the fight against Covid.

We salute you, FLCCC Team.

* * *

Pierre Kory,

I passed one of the videos featuring your work with ivermectin to a local doctor here in a town in west Texas with a population of 12,000. He has started prescribing ivermectin along with some other meds and my good friends got Covid. They are in their mid 60s, one being treated for prostate cancer. They have breezed through it with merely some upset stomach symptoms and a slight cough at night.

* * *

Happy New Year Dr. Kory.

I am an ER physician in Arizona. I contracted COVID on December 12 and it was horrific for me. I ended up needing almost 3 weeks to recover but was saved by many avenues. One of the key components was Ivermectin. It took me 2 rounds of ivermectin, but this drug cured me and allowed me to overcome COVID. My entire family contracted COVID and yes, they all took ivermectin and were just fine. Since then, several family/friends/church members have all received the multi drug therapy (ivermectin) and have truly been saved.

Even as an ER physician, I was struggling with panic and fear, but I was fortunate and lucky to find hope. I cannot imagine what so many Americans are feeling at home and dying at unnecessary rates as this cure is literally underneath our noses.

[We] have been given strict orders to not give ANY ivermectin, steroids, or even hydroxychloroquine. This entire fiasco is downright criminal. The ordeal in itself is a story for another day.

Thank you for being courageous to speak. Your testimony has saved lives, including my own.

This comment was from "Ex-Pharma Man" on our anniversary Substack post.

Thank you to all of you passionate, compassionate, and courageous doctors. I shared the protocol with more than 20 friends and family (a few that were on the verge of hospitalization) and every one of them responded with complete recovery within seven days of treatment. All of you are the epitome of what Hippocrates meant when he said, 'where there is love for medicine, there is a love for humanity.' Thank you all!!!

I have dozens more testimonials like these, and each and every one not only warms my heart, but together they are the fuel that keeps me going on the darkest of days.

CHAPTER FORTY-FOUR

After the End

*In one way we think a great deal too much of the atomic bomb.
"How are we to live in an atomic age?" I am tempted to reply:
"Why, as you would have lived in the sixteenth century when
the plague visited London almost every year, or as you would
have lived in a Viking age when raiders from Scandinavia
might land and cut your throat any night; or indeed, as you
are already living in an age of cancer, an age of syphilis, an age
of paralysis, an age of air raids, an age of railway accidents, an
age of motor accidents. If we are all going to be destroyed by an
atomic bomb, let that bomb when it comes find us doing sensible and human things—praying, working, teaching, reading,
listening to music, bathing the children, playing tennis, chat-
ting to our friends over a pint and a game of darts—not hud-
dled together like frightened sheep and thinking about bombs.
They may break our bodies (a microbe can do that) but they
need not dominate our minds."*

—C. S. Lewis

I began this book by introducing you to "old Pierre," the journal-devouring,
New York Times–worshiping, vaccine-loving physician who meant well
but was blind to what had become of the medical system he thought he
knew. Old Pierre is now at the tail end of a three-year journey of discovery;

a journey during which he lost three jobs and his academic and teaching career ended. Today he has state licensing boards and professional societies trying to discredit, delicense, and decertify his ass; the media dreams of destroying him. And if he could go back in time and do it all over again, he wouldn't change a single godforsaken thing.

Let me clarify: *Of course, if he had a cape and a magic wand and unlimited wishes, he'd overhaul the corrupt health care system and fire all of the fake journalists and fraudulent scientists and sprinkle ivermectin like fairy dust all around the world.* But he'd still stand up and fight back and call bullshit every chance he could get. He'd lose the jobs and suffer the insults and be roundly disappointed by humanity a hundred times over, because he knows with absolute certainty that he is on the right side of history.

Mark Twain said, "Whenever you find yourself on the side of the majority, it is time to pause and reflect." If only someone would mandate reading more Mark Twain.

Sometimes I wonder about my legacy, as well as those of the people who have been on the wrong side of this thing: the scientists and journalists who accepted large bribes and small ones; the doctors and nurses who looked away when people were suffering, or refused to treat unvaccinated patients, or who muttered, "I'm sorry, I'm just following orders," as they forbade families from being together in those precious, final moments of life; the pediatricians who are seeing their young patients suffer heart attacks and strokes and every other sudden syndrome under the sun and yet keep jabbing those tiny arms day in and day out. Can they look at themselves in the mirror? Can they meet their children's eyes without shame?

I can.

Daily I drown in patients amid the humanitarian catastrophe of vaccine injuries and deaths while so many former colleagues are gaslighting and dismissing injured patients and actively pushing the next round of boosters. I worry about the population effects of the sudden and massive drops in birth rates around the world. I'm disturbed by the palpable losses from the workforce via the disability and death of working age Americans. I fear— how is there not an actual word for *shake with terror on a cellular level?*—the long-term, still unknown effects of this genetic therapy experiment that was forced upon humanity.

As the father of three daughters, the silence of the world's OB-GYNs has been the most terrifying. There are hundreds of thousands of women suffering menstrual irregularities post vaccination, yet reproductive specialists continue to insist the vaccines are safe and effective in pregnancy (and otherwise). *They are neither.* Birth rates are dropping precipitously around the world. Miscarriages and birth defects have skyrocketed as described in a recent paper by Jim Thorpe,[1] a colleague and dear friend, and one of the few OB-GYN specialists in the US who has fought for the health of our mothers and daughters and our future. Are the rest of the OB-GYNs unwitting stooges or corporate fascists? Frankly, I don't care. Their silence is criminal.

Where is the outrage? I have felt as if I've had enough for all of us. I waited for fired-up liberal college students to rise up and rally against the jab mandates. I stood by hoping for musicians to write angry, impassioned songs full of insight and truth, denouncing the injustice of it all. The silence of the global pushback has been deafening, with rare exceptions including my new friends and allies Five Times August, Right Said Fred, Zubi, Pete Parada, Jimmy Levy, Jason Aldean, Eric Clapton (bravely outspoken since his own vaccine injury[2]), and a handful of others.

Instead, you heard A-list musicians and red-carpet celebrities issuing plaintive cries to *do the right thing and get the vaccine* (Pink, Michael Phelps, Jimmy Kimmel, Dolly Parton, Martha Stewart, Oprah Winfrey, Ryan Reynolds, Mariah Carey, and Amy Schumer, I'm looking at you). At the same time, dozens of their peers were suffering strokes and paralysis and dropping dead but don't you dare ask about their vaccine statuses. That's personal, private, HIPAA-protected information. Unless you wanted to attend nearly every US university, eat in a restaurant in New York City, or play in the US Open during the peak of Covid.

Socially, I rarely talk to any of my former medical colleagues and only a few of my old friends. Intellectually, I no longer trust journalists writing on any topic outside of fashion, entertainment, art, or travel. I read newspapers now just to see what "the other side" is up to. What's their latest storyline or messaging to get me to believe or act in a way that furthers the goals of whoever in power is pulling those strings? Armed with this understanding, I can do the opposite. And, I believe, so should you.

The government has lost my trust—and not just with regard to medicine. Climate change, the war in Ukraine, oil shortages, a banking collapse—whatever the new "narrative" is, I'm inclined to suspect it's usually and mostly false; willfully constructed not for the common good, but for the good of those who control the institutions of society. It's so bad that I am now even suspicious of the relentless refrain, "Ninety-eight percent of climate change scientists agree that CO_2 is the root cause of global warming." Do you know why? Because I heard that same "consensus" about ivermectin, HCQ, and the vaccines. It might be true; it may not be. I may have time to "do my own research" someday; I may not. Regardless, they blew it. I will never again believe a word they say without verifying it myself through exhaustive investigation.

For the record, I still believe that the vast, overwhelming majority of people are good. And I don't necessarily know who "they" are either; the evil forces responsible for the carnage I've seen. I certainly have a list of suspects; Big Pharma and Bill Gates and Klaus Schwab and the rest of the globalist brat pack are on there, but who is above them? The industrialists with more money than God? The Rockefellers, the Bilderbergs, BlackRock, the Illuminati? Who knows? All I know is that I was put on this planet at this time in history for a reason, and so were you. Or maybe we weren't. Maybe it's all just a crap shoot and *you get what you get and you don't throw a fit.* But I like to believe there's a reason. It galvanizes me.

My mother is from France and my father is from Hungary; technically he is Jewish because his mother was Jewish, although his father was Christian. Most of my grandmother's family were exterminated in the Holocaust. My father only survived by being taken in by a Christian uncle. My grandmother was one of the few survivors of Auschwitz, and she was an absolute badass. She was raised in a well-to-do family, was highly educated, and was considered a great beauty throughout the region of Transylvania. To my great-grandfather's ire, she married a poor, Christian painter, an act of rebellion for which her father promptly disowned and disinherited her. As if that weren't bad enough, she then proceeded to get a divorce. *Quelle horreur!* A wealthy, educated girl from a prominent Jewish family married a Christian painter and then *got a divorce.* In the 1920s. She may as well have opened her own speakeasy and announced she was enlisting in the Marines.

After she emigrated to the US, my grandmother got a job as an assistant librarian at Columbia University in New York City. With the free tuition that came with the job, she ended up getting a PhD in French literature and eventually became a professor at Hofstra University in Long Island, NY, where I grew up. She lived to be eighty-nine and her badassery never waned.

I loved my grandmother and got very close to her in my twenties. We'd take long walks together and I would record our conversations, as her life was fascinating and, I felt, filled with incredible lessons. One day, with my tape recorder in hand, I asked her, "Anyu [the word for mother in

Hungarian], why do you think you survived Auschwitz when so few others did?" (Among the executed was her second husband and the love of her life.) Her reply gave me chills. "I think it is because I would look the guards in the eye, showing them that I was not scared of them," she told me. "I think they respected that."

I know I respect the hell out of that. And I like to think I have a little bit of her badassery in me. I may be sad about the state of medicine and nervous about what the future holds for mankind, but I'm not afraid to look the guards right in the fecking eye.

Who's with me?

Notes

Chapter One

[1] https://en.wikipedia.org/wiki/Regulatory_capture.

Chapter Two

[1] https://www.med.ubc.ca/news/sepsis-leading-cause-of-death-worldwide/.
[2] https://neuroimmune.org/.

Chapter Six

[1] https://www.cancer.gov/publications/dictionaries/cancer-terms/def/randomized-clinical-trial.

Chapter Nine

[1] https://www.nytimes.com/2020/08/05/magazine/covid-drug-wars-doctors.html.
[2] https://jamanetwork.com/journals/jama/fullarticle/2752063.
[3] https://pubmed.ncbi.nlm.nih.gov/34833042/.

Chapter Ten

[1] https://covid19criticalcare.com/dr-kory-testifies-before-u-s-senate-committee-on-homeland-security-and-governmental-affairs-on-treating-covid-19-may-6-2020/.
[2] https://odysee.com/@FrontlineCovid19CriticalCareAlliance:c/Dr.-Pierre-Kory-FLCCC-Alliance-testifies-to-senate-committee-about-I-MASK-incl.-the-following-QA-part-490351508:3.

Chapter Twelve

[1] https://www.youtube.com/watch?v=k9GYTc53r2o&t=2s.
[2] https://bmjopenrespres.bmj.com/content/7/1/e000724.

Chapter Thirteen

1   https://www.nytimes.com/2020/05/29/health/coronavirus-transmission-dose.html?referringSource=articleShare.
2   https://www.pnas.org/doi/10.1073/pnas.2009637117.
3   https://masks4all.co/.
4   https://www.preprints.org/manuscript/202004.0203/v1.
5   https://www.cdc.gov/mmwr/volumes/69/wr/mm6919e6.htm.
6   https://www.journalofinfection.com/article/S0163-4453(20)30117-1/fulltext.
7   https://www.koreaherald.com/view.php?ud=20200512000586.
8   https://www.bbc.com/news/53137613.
9   https://www.usatoday.com/story/opinion/2020/07/01/slow-covid-19-more-americans-need-wear-n-95-masks-indoors-column/3278779001/.

Chapter Fourteen

1   https://pierrekory.substack.com/p/expert-witness-testimony-of-the-george.

Chapter Fifteen

1   https://www.today.com/health/viral-photo-shows-icu-doctor-embracing-covid-19-patient-t201598.
2   https://web.archive.org/web/20220809070653/http://www.raumfisch.de/sign/?language=en.
3   https://covid19criticalcare.com/a-doctor-needed-a-life-flight-to-a-texas-hospital-to-get-a-pill-that-would-save-his-life-feb-12-2021/.
4   https://www.drbeen.com/.
5   https://www.jocmr.org/index.php/JOCMR/article/view/4658/25893530.
6   https://www.medscape.com/viewarticle/942995.

Chapter Sixteen

1   https://www.nejm.org/doi/full/10.1056/nejmoa2016638.
2   https://osf.io/bjx76.
3   https://pubmed.ncbi.nlm.nih.gov/26086943/.
4   https://www.researchgate.net/publication/344469305_Real-World_Evidence_The_Case_of_Peru_Causality_between_Ivermectin_and_COVID-19_Infection_Fatality_Rate.

Chapter Seventeen

1   https://covid19criticalcare.com/ivermectin/.
2   https://covid19criticalcare.com/flccc-alliance-news-conference-medical-evidence-of-ivermectin-effectively-prevent-treat-covid-19/.
3   https://changemakerfdn.org/.

Chapter Eighteen

[1] https://www.merck.com/company-overview/history/.
[2] https://covid19criticalcare.com/dr-pierre-kory-flccc-alliance-testifies-to-u-s-senate-committee-about-i-mask/.
[3] https://www.hsgac.senate.gov/wp-content/uploads/imo/media/doc/Testimony-Kory-2020-12-08.pdf.

Chapter Nineteen

[1] https://c19ivm.org/.
[2] https://c19ivm.org/meta.html.
[3] https://cdn-www.lanacionpy.arcpublishing.com/politica/2020/09/01/gobernador-distribuira-ivermectina-para-combate-al-covid-19-en-alto-parana/.
[4] www.pagina16.com.ar/ivermectina-brindan-resultados-parciales-de-monitoreo-en-el-uso-ampliado-en-pacientes-positivos/.
[5] http://www.pharmabaires.com/1767-salta-y-corrientes-adoptan-ivermectina-en-sus-protocolos-covid.html.
[6] https://www.cureus.com/articles/82162-ivermectin-prophylaxis-used-for-covid-19-a-citywide-prospective-observational-study-of-223128-subjects-using-propensity-score-matching#!/.
[7] https://www.diariodelsur.com.mx/local/en-la-ultima-semana-han-disminuido-el-numero-de-casos-covid-19-secretaria-de-salud-pandemia-conferencia-de-prensa-clinicas-comida-chatarra-5411551.html.
[8] https://papers.ssrn.com/sol3/papers.cfm?abstract_id=3765018.
[9] https://www.antibiotics.or.jp/wp-content/uploads/74-1_44-95.pdf.
[10] https://c19early.org/ace.
[11] https://c19early.org/cpmeta.html.
[12] https://www.covid19treatmentguidelines.nih.gov/overview/prevention-of-sars-cov-2/.
[13] https://www.medincell.com/wp-content/uploads/2023/01/PR-results-TTG-VF-EN.pdf.

Chapter Twenty-One

[1] https://en.wikipedia.org/wiki/List_of_largest_pharmaceutical_settlements.
[2] Peter Gøtzsche, *Deadly Medicines and Organised Crime* (Boca Raton: CRC Press, 2013).
[3] https://apnews.com/article/health-north-america-us-news-ap-top-news-dance-82f638d6dfcf4193ad28ddf0e65897e1.
[4] Peter Rost, *The Whistleblower: Confessions of a Healthcare Hitman* (New York: Soft Skull Press, 2006).
[5] https://www.investopedia.com/investing/which-industry-spends-most-lobbying-antm-so/.

Chapter Twenty-Two

[1] https://www.trialsitenews.com/a/trialsite-news-original-documentary-in-peru-about-ivermectin-and-covid-19.

2 https://www.fda.gov/safety/medical-product-safety-information/ivermectin-intended-animals-letter-stakeholders-do-not-use-humans-treatment-covid-19.

3 https://www.sciencedirect.com/science/article/pii/S0166354220302199.

4 https://www.sciencedirect.com/science/article/pii/S0166354220302199#bib3.

5 https://www.certara.com/pressrelease/pfizer-deploys-certaras-d360-for-scientific-data-access-and-analysis/.

6 https://www.sciencedirect.com/science/article/pii/S015196382030627X?via%3Dihub.

Chapter Twenty-Three

1 https://www.ncbi.nlm.nih.gov/pmc/articles/PMC8088823/.

2 https://newrepublic.com/article/162000/bill-gates-impeded-global-access-covid-vaccines.

3 https://www.youtube.com/watch?v=M8RMBa1UfsE.

4 https://bonsens.info/dr-andrew-hill-sur-livermectine/.

5 https://www.oraclefilms.com/alettertoandrewhill.

6 https://bird-group.org/.

7 https://www.20minutes.fr/societe/2935679-20201219-coronavirus-non-ivermectine-medicament-tres-efficace-contre-maladie.

8 https://twitter.com/DgCostagliola/status/1371085649562976256.

9 https://covid19criticalcare.com/videos-and-press/flccc-releases/flccc-alliance-statement-on-the-irregular-actions-of-public-health-agencies-and-the-widespread-disinformation-campaign-against-ivermectin/.

10 https://www.who.int/news-room/feature-stories/detail/who-advises-that-ivermectin-only-be-used-to-treat-covid-19-within-clinical-trials.

Chapter Twenty-Four

1 https://www.ivermectin.africa/2021/01/20/dr-andrew-hall-offers-latest-meta-analysis-data-on-ivermectin-to-sa-docters-on-20-jan-2020/.

2 https://academic.oup.com/ofid/article/8/11/ofab358/6316214?login=false.

3 https://www.cbsnews.com/news/merck-created-hit-list-to-destroy-neutralize-or-discredit-dissenting-doctors/.

4 https://grftr.news/why-was-a-major-study-on-ivermectin-for-covid-19-just-retracted/.

5 https://www.theguardian.com/science/2021/jul/16/huge-study-supporting-ivermectin-as-covid-treatment-withdrawn-over-ethical-concerns.

6 https://c19ivm.org/meta.html#tp.

7 https://doyourownresearch.substack.com/p/is-ivermectin-literature-particularly.

8 https://twitter.com/halfbakedopo/status/1632916546707398656/photo/2.

9 https://blogs.bmj.com/bmj/2021/07/05/time-to-assume-that-health-research-is-fraudulent-until-proved-otherwise/.

10 https://doyourownresearch.substack.com/p/is-ivermectin-literature-particularly.

11 https://geneticliteracyproject.org/2019/02/18/despite-controversial-new-study-theres-still-no-evidence-monsantos-roundup-causes-cancer-epidemiologist-says/.

12 https://www.e-epih.org/journal/view.php?doi=10.4178/epih.e2016014.

13 https://i-base.info/htb/40959.

Chapter Twenty-Five

[1] https://doyourownresearch.substack.com/p/activ-6-and-together-bear-strangely.
[2] https://doyourownresearch.substack.com/p/did-use-of-ivermectin-in-latin-america.
[3] https://doyourownresearch.substack.com/p/activ-6-dosing-and-timing-a-fox-in.
[4] https://doyourownresearch.substack.com/p/the-story-of-a-real-activ-6-patient.
[5] https://doyourownresearch.substack.com/p/activ-6-trial-ivermectin-scientists.
[6] https://c19ivm.org/togetherivm.html, https://c19ivm.org/activ6ivm.html, https://c19ivm.org/lim.html, https://c19ivm.org/lopezmedina.html.
[7] https://childrenshealthdefense.org/defender/health-officials-vaccines-cure-all-cdc-rochelle-walensky/.

Chapter Twenty-Six

[1] https://doyourownresearch.substack.com/p/together-trial-on-ivermectin-did.
[2] https://doyourownresearch.substack.com/p/together-trial-impossible-numbers.
[3] https://doyourownresearch.substack.com/p/together-trial-impossible-numbers.
[4] https://ir.eigerbio.com/news-releases/news-release-details/eiger-biopharmaceuticals-provides-update-status-planned.
[5] Marcia Angell, *The Truth About Drug Companies: How They Deceive Us and What to Do About It* (New York: Random House, 2004).

Chapter Twenty-Seven

[1] https://onlinelibrary.wiley.com/doi/10.1111/eci.12473.
[2] https://www.the-scientist.com/news-opinion/frontiers-pulls-special-covid-19-issue-after-content-dispute-68721.
[3] https://www.ncbi.nlm.nih.gov/pmc/articles/PMC8248252/.
[4] https://socopen.org/2022/02/04/on-withdrawing-ivermectin-and-the-odds-of-hospitalization-due-to-covid-19-by-merino-et-al/.
[5] https://covid19criticalcare.com/treatment-protocols/math-covid-hospital-treatment/.
[6] https://doyourownresearch.substack.com/p/academic-narrative-enforcers-the.
[7] https://www.the-scientist.com/news-opinion/the-top-retractions-of-2021-69533.

Chapter Twenty-Eight

[1] https://medicalpressopenaccess.com/upload/1605709669_1007.pdf.
[2] https://www.jcdr.net/articles/PDF/14529/46795_CE%5BRa%5D_F(Sh)_PF1(SY_OM)_PFA_(OM)_PN(KM).pdf.
[3] https://www.medrxiv.org/.
[4] https://www.forbes.com/sites/angelauyeung/2020/04/03/a-bill-gates-backed-accelerator-for-covid-19-coronavirus-therapeutics-treatment-partners-with-madonna-and-mark-zuckerbergs-chan-zuckerberg-initiative/?sh=820961d8067a.

Chapter Twenty-Nine

[1] https://apnews.com/article/fact-checking-afs:Content:9768999400.

Chapter Thirty-One

[1] https://www.theblaze.com/news/review-the-federal-government-paid-media-companies-to-advertise-for-the-vaccines.

[2] https://www.documentcloud.org/documents/23180971-senator-rand-paul-letter-to-cdc-on-weber-shandwick.

[3] https://www.bbc.com/mediacentre/2020/trusted-news-initiative-vaccine-disinformation.

[4] https://support.google.com/youtube/answer/9891785?hl=en.

[5] https://www.forbes.com/sites/brucelee/2021/06/12/youtube-suspends-republican-senator-ron-johnsons-account-for-violating-covid-19-policy/?sh=74cbda0c46eb.

[6] https://support.google.com/youtube/answer/9891785?hl=en.

[7] https://www.docubay.com/trust-who-2138.

[8] https://cyber.fsi.stanford.edu/content/virality-project.

[9] https://twitter.com/mtaibbi/status/1636729182712483842.

[10] https://nypost.com/2023/03/09/matt-taibbi-eviscerates-twitter-in-congressional-hearing/.

[11] Heather Heying and Bret Weinstein, *A Hunter-Gatherer's Guide to the 21st Century: Evolution and the Challenges of Modern Life* (New York: Portfolio, 2021).

[12] https://www.mountainhomemag.com/2021/05/01/356270/the-drug-that-cracked-covid.

[13] https://open.spotify.com/episode/7uVXKgE6eLJKMXkETwcw0D.

[14] https://en.wikipedia.org/wiki/Steve_Kirsch.

[15] https://www.vacsafety.org/.

[16] https://www.theepochtimes.com/cdc-partners-with-social-and-behavior-change-initiative-to-silence-vaccine-hesitancy_5175172.html.

[17] https://www.publicgoodprojects.org/work/infectious-diseases-and-vaccines.

[18] https://www.cdcfoundation.org/FY2021/donors?group=corp

Chapter Thirty-Three

[1] https://rescue.substack.com/p/horse-bleep-how-4-calls-on-animal.

[2] https://www.cnn.com/videos/health/2021/08/29/dr-anthony-fauci-ivermectin-covid-19-sotu-vpx.cnn.

[3] https://www.csis.org/analysis/conversation-dr-anthony-fauci-antiviral-program-pandemics.

Chapter Thirty-Four

[1] https://www.cbsnews.com/news/merck-created-hit-list-to-destroy-neutralize-or-discredit-dissenting-doctors/.

[2] https://www.the-scientist.com/the-nutshell/merck-published-fake-journal-44190.

[3] https://www.finance.senate.gov/imo/media/doc/111804dgtest.pdf.

[4] https://www.businessinsider.com/why-ivermectin-being-used-treat-Covid-2-doctors-leading-charge-2021-9.

[5] https://www.vice.com/en/article/wxda8q/the-ivermectin-guys-whole-thing-has-really-fallen-apart.

6   https://abcnews.go.com/US/ivermectin-condemned-experts-Covid-19-treatment-continues-easily/story?id=83178790.

7   https://www.scientificamerican.com/article/fringe-doctors-groups-promote-ivermectin-for-covid-despite-a-lack-of-evidence/.

8   https://www.statnews.com/2022/07/26/ivermectin-has-become-a-popular-treatment-for-long-covid-with-a-push-from-doctors-with-ties-to-right-wing-political-groups/.

9   https://www.huffpost.com/entry/ivermectin-compounding-pharmacies-covid_n_617c4 864e4b0931432187f71.

10 https://www.medpagetoday.com/special-reports/exclusives/102183.

11 https://www.cbc.ca/news/canada/calgary/coronavirus-cure-claim-1.5506187.

12 https://www.medpagetoday.com/special-reports/exclusives/93936.

13 https://www.mckinneyfamilymed.com/dr-mccullough

14 https://medicospelavidacovid19.com.br/.

15 https://www.biznews.com/health/2022/09/29/jackie-stone.

16 May 9, 1933: Helen Keller's Searing Letter to the Nazis About Censorship and the Inextinguishable Freedom of Ideas – The Marginalian.

Chapter Thirty-Five

1   https://www.skirsch.com/covid/TysonIvermectin.pdf.

Chapter Thirty-Six

1   https://www.mountainhomemag.com/2021/05/01/356270/the-drug-that-cracked-covid.

2   https://www.instagram.com/reel/Ck1n6g5smlL/?__coig_restricted=1.

Chapter Thirty-Eight

1   https://zeenews.india.com/india/yogi-government-cracks-whip-on-775-corrupt-officials-since-it-takes-charge-in-uttar-pradesh-2308884.html.

2   http://dgmhup.gov.in/documents/AdvisoryontheuseofHydroxychloroquinas prophylaxisforSARSCoV2infection.pdf (Note: To view, you must set your VPN to India).

3   https://indianexpress.com/article/india/up-new-protocol-ivermectin-to-replace-hcq-in-treatment-of-covid-patients-6545236/.

4   https://indianexpress.com/article/opinion/columns/coronavirus-pandemic-covid-vaccine-tracker-uttar-pradesh-7107756/.

5   https://indianexpress.com/article/cities/lucknow/uttar-pradesh-government-says-ivermectin-helped-to-keep-deaths-low-7311786/.

6   https://www.ncbi.nlm.nih.gov/pmc/articles/PMC7485636/.

7   https://www.who.int/docs/default-source/wrindia/document/uttar-pradesh_case-study.pdf?sfvrsn=b4c0f723_4.

8   https://www.amarujala.com/uttar-pradesh/agra/corona-virus-infected-person-contect-take-ivermectin-now-in-up?pageId=2.

9   https://www.livemint.com/news/india/uttar-pradesh-covid-tests-at-airport-railway-bus-stations-for-people-coming-from-states-with-high-caseload-11616545854641.html.

10 https://sundayguardianlive.com/news/learnings-uttar-pradeshs-management-second-covid-19-wave.

11 https://www.india.com/hindi-news/uttar-pradesh/medicine-for-corona-virus-patients-people-do-treatment-without-waiting-of-rtpcr-report-up-government-released-7-medicines-list-4593321/.

12 https://www.india.com/uttar-pradesh/dont-turn-away-patients-will-bear-cost-of-treatment-for-covid-patients-up-govt-tells-hospitals-4614067/.

13 https://indianexpress.com/article/cities/lucknow/uttar-pradesh-government-says-ivermectin-helped-to-keep-deaths-low-7311786/.

14 https://www.hindustantimes.com/india-news/how-uttar-pradesh-turned-the-tide-in-fight-against-covid19-101621590095721.html.

15 https://timesofindia.indiatimes.com/city/lucknow/65-of-record-27-lakh-vaccines-given-to-uttar-pradesh-rural-population/articleshow/85058420.cms.

16 https://www.who.int/india/news/feature-stories/detail/uttar-pradesh-going-the-last-mile-to-stop-covid-19.

17 https://www.india.com/news/india/up-now-prepares-for-third-covid-wave-readies-special-medicine-kits-for-kids-report-4727789/.

18 https://www.hindustantimes.com/cities/lucknow-news/33-districts-in-uttar-pradesh-are-now-covid-free-state-govt-101631267966925.html.

19 https://www.indiatoday.in/coronavirus-outbreak/story/uttar-pradesh-districts-covid-free-cases-deaths-1847365-2021-08-31.

20 https://drive.google.com/file/d/12Hp6LPYBXmrguADWVbc8fwYO4pGTkkvA/view.

Chapter Thirty-Nine

1 https://www.mohfw.gov.in/pdf/Revisedadvisoryontheuseofhydroxychloroquineas prophylaxisforSARSCOVID19infection.pdf.

2 https://health.economictimes.indiatimes.com/news/pharma/why-icmr-continues-to-stand-firm-on-using-hydroxychloroquine-as-prophylaxis/76172274.

3 https://qz.com/india/1968541/india-distributed-over-100-million-hcq-tablets-for-covid-19.

4 https://www.news18.com/news/india/india-boosts-output-of-hydroxychloroquine-drug-exported-to-97-countries-amid-pandemic-says-health-min-2610609.html.

5 https://health.uk.gov.in/files/RevisedguidelinesforHomeIsolationofmildasymptomatic COVID19cases_1.pdf.

6 https://twitter.com/search?q=%40doctorsoumya%20WHO%20mild%20covid%20 home%20care%20bundle&src=typed_query&f=top

7 https://twitter.com/drasmalhi/status/1392072320483696641

8 https://swarajyamag.com/news-brief/indian-bar-association-vs-who-chief-scientist-soumya-swaminathan-does-ivermectin-help-covid-patients

9 indiatoday.in/india/story/ayushman-bharat-digital-mission-to-ensure-equitable-healthcare-says-bill-gates-pm-modi-thanks-him-1858875-2021-09-29.

10 https://www.indiatoday.in/coronavirus-outbreak/story/why-hcq-ivermectin-dropped-india-covid-treatment-protocol-1857306-2021-09-25.

11 https://cdn-www.lanacionpy.arcpublishing.com/politica/2020/09/01/gobernador-distribuira-ivermectina-para-combate-al-covid-19-en-alto-parana/.

12 www.pagina16.com.ar/ivermectina-brindan-resultados-parciales-de-monitoreo-en-el-uso-ampliado-en-pacientes-positivos/.

13 http://www.pharmabaires.com/1767-salta-y-corrientes-adoptan-ivermectina-en-sus-protocolos-covid.html.

14 https://www.cureus.com/articles/82162-ivermectin-prophylaxis-used-for-covid-19-a-citywide-prospective-observational-study-of-223128-subjects-using-propensity-score-matching#!/.

15 https://www.diariodelsur.com.mx/local/en-la-ultima-semana-han-disminuido-el-numero-de-casos-covid-19-secretaria-de-salud-pandemia-conferencia-de-prensa-clinicas-comida-chatarra-5411551.html.

16 https://papers.ssrn.com/sol3/papers.cfm?abstract_id=3765018.

Chapter Forty

1 https://www.nejm.org/doi/10.1056/NEJMoa2034577.

2 https://bmcinfectdis.biomedcentral.com/articles/10.1186/s12879-023-07998-3.

3 https://rescue.substack.com/p/the-missing-babies-of-europe

Chapter Forty-One

1 https://thepostmillennial.com/youtube-removes-joe-rogan-interview-with-dr-robert-malone-from-youtube.

2 https://www.usatoday.com/story/tech/2021/06/28/google-search-adds-warning-rapidly-evolving-results/5370360001/.

3 https://newsrescue.com/google-quickly-resets-screens-results-for-mass-formation-psychosis-after-bombshell-dr-malone-joe-rogan-interview/.

4 https://wtop.com/dc/2021/12/bowser-proof-of-vaccination-to-be-required-at-dc-restaurants-bars/.

5 https://www.forbes.com/sites/brucelee/2022/03/05/covid-19-freedom-protest-blocks-dc-area-traffic-truckersconvoy2022-contrasts-this-with-ukraine-crisis/?sh=744941f87950.

6 https://rumble.com/vwyb1p-1080p-senators-ron-johnson-and-ted-cruze-meet-with-the-peoples-convoy.html.

Chapter Forty-Two

1 https://www.realclearmarkets.com/articles/2022/10/25/californias_dangerous_strides_to_criminalize_medical_disagreement_860928.html.

2 https://covid19criticalcare.com/wp-content/uploads/2023/02/I-RECOVER-Post-Vaccine-2023-03-11.pdf.

3 https://pierrekory.substack.com/.

4 https://worldcouncilforhealth.org/news/launch-with-publication-of-at-home-treatment-guide/.

5 https://www.facebook.com/771203558/posts/pfbid0Q6beanUDocR8zN7VBNEGK6rqnQq9AndZUzbdzwHop6P8aPyppJ5QkHjcQCU7oygrl/?mibextid=ykz3hl.

6 https://www.youtube.com/watch?v=9jMONZMuS2U

7 https://twitter.com/justincbzz/status/1446780274201833476

8 https://mspsss.org.ua/index.php/journal/article/view/342.

9 https://www.clarkcountytoday.com/news/17000-doctors-call-for-end-to-covid-19-emergency/.

10   https://odysee.com/@FrontlineCovid19CriticalCareAlliance:c/Medical-Freedom-Panel-Ohio-State-Capitol:9?lid=61530282b09fe7b2623f851d208336934d54e64b.

11   https://odysee.com/@ysambre_engag%C3%A9:5/Dr-Pierre-Kory:3.

Chapter Forty-Four

1   https://www.jpands.org/vol28no1/thorp.pdf.

2   https://www.rollingstone.com/music/music-news/eric-clapton-disastrous-vaccine-propaganda-1170264/